False twist textured yarns

The Textile Institute and Woodhead Publishing

The Textile Institute is a unique organisation in textiles, clothing and footwear. Incorporated in England by a Royal Charter granted in 1925, the Institute has individual and corporate members in over 90 countries. The aim of the Institute is to facilitate learning, recognise achievement, reward excellence and disseminate information within the global textiles, clothing and footwear industries.

Historically, The Textile Institute has published books of interest to its members and the textile industry. To maintain this policy, the Institute has entered into partnership with Woodhead Publishing Limited to ensure that Institute members and the textile industry continue to have access to high calibre titles on textile science and technology.

Most Woodhead titles on textiles are now published in collaboration with The Textile Institute. Through this arrangement, the Institute provides an Editorial Board which advises Woodhead on appropriate titles for future publication and suggests possible editors and authors for these books. Each book published under this arrangement carries the Institute's logo.

Woodhead books published in collaboration with The Textile Institute are offered to Textile Institute members at a substantial discount. These books, together with those published by The Textile Institute that are still in print, are offered on the Woodhead web site at: www.woodheadpublishing.com. Textile Institute books still in print are also available directly from the Institute's web site at: www.textileinstitutebooks.com.

A list of Woodhead books on textiles science and technology, most of which have been published in collaboration with the Textile Institute, can be found towards the end of the contents pages.

Woodhead Publishing Series in Textiles: Number 129

False twist textured yarns

Principles, processes and applications

C. Atkinson

The Textile Institute

WOODHEAD PUBLISHING

Oxford Cambridge Philadelphia New Delhi

Published by Woodhead Publishing Limited in association with The Textile Institute
Woodhead Publishing Limited, 80 High Street, Sawston,
Cambridge CB22 3HJ, UK
www.woodheadpublishing.com
www.woodheadpublishingonline.com

Woodhead Publishing, 1518 Walnut Street, Suite 1100, Philadelphia,
PA 19102-3406, USA

Woodhead Publishing India Private Limited, G-2, Vardaan House,
7/28 Ansari Road, Daryaganj, New Delhi – 110002, India
www.woodheadpublishingindia.com

First published 2012, Woodhead Publishing Limited

British Library Cataloguing in Publication Data
A catalogue record for this book is available from the British Library.

Library of Congress Control Number: 2012931667

ISBN 978-1-84569-933-8 (print)
ISBN 978-0-85709-559-6 (online)
ISSN 2042-0803 Woodhead Publishing Series in Textiles (print)
ISSN 2042-0811 Woodhead Publishing Series in Textiles (online)

The publisher's policy is to use permanent paper from mills that operate a sustainable forestry policy, and which has been manufactured from pulp which is processed using acid-free and elemental chlorine-free practices. Furthermore, the publisher ensures that the text paper and cover board used have met acceptable environmental accreditation standards.

Typeset by Toppan Best-set Premedia Limited, Hong Kong
Printed by TJI Digital, Padstow, Cornwall, UK

Contents

Woodhead Publishing Series in Textiles

1 **Watson's textile design and colour Seventh edition**
Edited by Z. Grosicki

2 **Watson's advanced textile design**
Edited by Z. Grosicki

3 **Weaving Second edition**
P. R. Lord and M. H. Mohamed

4 **Handbook of textile fibres Vol 1: Natural fibres**
J. Gordon Cook

5 **Handbook of textile fibres Vol 2: Man-made fibres**
J. Gordon Cook

6 **Recycling textile and plastic waste**
Edited by A. R. Horrocks

7 **New fibers Second edition**
T. Hongu and G. O. Phillips

8 **Atlas of fibre fracture and damage to textiles Second edition**
J. W. S. Hearle, B. Lomas and W. D. Cooke

9 **Ecotextile '98**
Edited by A. R. Horrocks

10 **Physical testing of textiles**
B. P. Saville

11 **Geometric symmetry in patterns and tilings**
C. E. Horne

12 **Handbook of technical textiles**
Edited by A. R. Horrocks and S. C. Anand

13 **Textiles in automotive engineering**
W. Fung and J. M. Hardcastle

14 **Handbook of textile design**
J. Wilson

15 **High-performance fibres**
Edited by J. W. S. Hearle

16 **Knitting technology Third edition**
D. J. Spencer

17 **Medical textiles**
Edited by S. C. Anand

18 **Regenerated cellulose fibres**
Edited by C. Woodings

19 **Silk, mohair, cashmere and other luxury fibres**
Edited by R. R. Franck

20 **Smart fibres, fabrics and clothing**
Edited by X. M. Tao

21 **Yarn texturing technology**
J. W. S. Hearle, L. Hollick and D. K. Wilson

22 **Encyclopedia of textile finishing**
H-K. Rouette

23 **Coated and laminated textiles**
W. Fung

24 **Fancy yarns**
R. H. Gong and R. M. Wright

25 **Wool: Science and technology**
Edited by W. S. Simpson and G. Crawshaw

26 **Dictionary of textile finishing**
H-K. Rouette

27 **Environmental impact of textiles**
K. Slater

28 **Handbook of yarn production**
P. R. Lord

29 **Textile processing with enzymes**
Edited by A. Cavaco-Paulo and G. Gübitz

30 **The China and Hong Kong denim industry**
Y. Li, L. Yao and K. W. Yeung

31 **The World Trade Organization and international denim trading**
Y. Li, Y. Shen, L. Yao and E. Newton

32 **Chemical finishing of textiles**
W. D. Schindler and P. J. Hauser

33 **Clothing appearance and fit**
J. Fan, W. Yu and L. Hunter

34 **Handbook of fibre rope technology**
H. A. McKenna, J. W. S. Hearle and N. O'Hear

35 **Structure and mechanics of woven fabrics**
J. Hu

36 **Synthetic fibres: nylon, polyester, acrylic, polyolefin**
Edited by J. E. McIntyre

37 **Woollen and worsted woven fabric design**
E. G. Gilligan

38 **Analytical electrochemistry in textiles**
P. Westbroek, G. Priniotakis and P. Kiekens

39 **Bast and other plant fibres**
R. R. Franck

40 **Chemical testing of textiles**
Edited by Q. Fan

41 **Design and manufacture of textile composites**
Edited by A. C. Long

42 **Effect of mechanical and physical properties on fabric hand**
Edited by Hassan M. Behery

43 **New millennium fibers**
T. Hongu, M. Takigami and G. O. Phillips

44 **Textiles for protection**
Edited by R. A. Scott

45 **Textiles in sport**
Edited by R. Shishoo

46 **Wearable electronics and photonics**
Edited by X. M. Tao

47 **Biodegradable and sustainable fibres**
Edited by R. S. Blackburn

48 **Medical textiles and biomaterials for healthcare**
Edited by S. C. Anand, M. Miraftab, S. Rajendran and J. F. Kennedy

49 **Total colour management in textiles**
Edited by J. Xin

50 **Recycling in textiles**
Edited by Y. Wang

51 **Clothing biosensory engineering**
Y. Li and A. S. W. Wong

52 **Biomechanical engineering of textiles and clothing**
Edited by Y. Li and D. X-Q. Dai

53 **Digital printing of textiles**
Edited by H. Ujiie

54 **Intelligent textiles and clothing**
Edited by H. R. Mattila

55 **Innovation and technology of women's intimate apparel**
W. Yu, J. Fan, S. C. Harlock and S. P. Ng

56 **Thermal and moisture transport in fibrous materials**
Edited by N. Pan and P. Gibson

57 **Geosynthetics in civil engineering**
Edited by R. W. Sarsby

58 **Handbook of nonwovens**
Edited by S. Russell

59 **Cotton: Science and technology**
Edited by S. Gordon and Y-L. Hsieh

60 **Ecotextiles**
Edited by M. Miraftab and A. R. Horrocks

61 **Composite forming technologies**
Edited by A. C. Long

62 **Plasma technology for textiles**
Edited by R. Shishoo

63 **Smart textiles for medicine and healthcare**
Edited by L. Van Langenhove

64 **Sizing in clothing**
Edited by S. Ashdown

65 **Shape memory polymers and textiles**
J. Hu

66 **Environmental aspects of textile dyeing**
Edited by R. Christie

67 **Nanofibers and nanotechnology in textiles**
Edited by P. Brown and K. Stevens

68 **Physical properties of textile fibres Fourth edition**
W. E. Morton and J. W. S. Hearle

69 **Advances in apparel production**
Edited by C. Fairhurst

70 **Advances in fire retardant materials**
Edited by A. R. Horrocks and D. Price

71 **Polyesters and polyamides**
Edited by B. L. Deopura, R. Alagirusamy, M. Joshi and B. S. Gupta

72 **Advances in wool technology**
Edited by N. A. G. Johnson and I. Russell

73 **Military textiles**
Edited by E. Wilusz

74 **3D fibrous assemblies: Properties, applications and modelling of three-dimensional textile structures**
J. Hu

75 **Medical and healthcare textiles**
Edited by S. C. Anand, J. F. Kennedy, M. Miraftab and S. Rajendran

76 **Fabric testing**
Edited by J. Hu

77 **Biologically inspired textiles**
Edited by A. Abbott and M. Ellison

78 **Friction in textile materials**
Edited by B. S. Gupta

79 **Textile advances in the automotive industry**
Edited by R. Shishoo

80 **Structure and mechanics of textile fibre assemblies**
Edited by P. Schwartz

81 **Engineering textiles: Integrating the design and manufacture of textile products**
Edited by Y. E. El-Mogahzy

82 **Polyolefin fibres: Industrial and medical applications**
Edited by S. C. O. Ugbolue

83 **Smart clothes and wearable technology**
Edited by J. McCann and D. Bryson

84 **Identification of textile fibres**
Edited by M. Houck

85 **Advanced textiles for wound care**
Edited by S. Rajendran

86 **Fatigue failure of textile fibres**
Edited by M. Miraftab

87 **Advances in carpet technology**
Edited by K. Goswami

88 **Handbook of textile fibre structure Volume 1 and Volume 2**
Edited by S. J. Eichhorn, J. W. S. Hearle, M. Jaffe and T. Kikutani

89 **Advances in knitting technology**
Edited by K-F. Au

90 **Smart textile coatings and laminates**
Edited by W. C. Smith

91 **Handbook of tensile properties of textile and technical fibres**
Edited by A. R. Bunsell

92 **Interior textiles: Design and developments**
Edited by T. Rowe

93 **Textiles for cold weather apparel**
Edited by J. T. Williams

94 **Modelling and predicting textile behaviour**
Edited by X. Chen

95 **Textiles, polymers and composites for buildings**
Edited by G. Pohl

96 **Engineering apparel fabrics and garments**
J. Fan and L. Hunter

97 **Surface modification of textiles**
Edited by Q. Wei

98 **Sustainable textiles**
Edited by R. S. Blackburn

99 **Advances in yarn spinning technology**
Edited by C. A. Lawrence

100 **Handbook of medical textiles**
Edited by V. T. Bartels

101 **Technical textile yarns**
Edited by R. Alagirusamy and A. Das

102 **Applications of nonwovens in technical textiles**
Edited by R. A. Chapman

103 **Colour measurement: Principles, advances and industrial applications**
Edited by M. L. Gulrajani

104 **Fibrous and composite materials for civil engineering applications**
Edited by R. Fangueiro

105 **New product development in textiles: Innovation and production**
Edited by L. Horne

106 **Improving comfort in clothing**
Edited by G. Song

107 **Advances in textile biotechnology**
Edited by V. A. Nierstrasz and A. Cavaco-Paulo

108 **Textiles for hygiene and infection control**
Edited by B. McCarthy

109 **Nanofunctional textiles**
Edited by Y. Li

110 **Joining textiles: Principles and applications**
Edited by I. Jones and G. Stylios

111 **Soft computing in textile engineering**
Edited by A. Majumdar

112 **Textile design**
Edited by A. Briggs-Goode and K. Townsend

113 **Biotextiles as medical implants**
Edited by M. King and B. Gupta

114 **Textile thermal bioengineering**
Edited by Y. Li

115 **Woven textile structure**
B. K. Behera and P. K. Hari

116 **Handbook of textile and industrial dyeing. Volume 1: Principles, processes and types of dyes**
Edited by M. Clark

117 **Handbook of textile and industrial dyeing. Volume 2: Applications of dyes**
Edited by M. Clark

118 **Handbook of natural fibres. Volume 1: Types, properties and factors affecting breeding and cultivation**
Edited by R. Kozlowski

119 **Handbook of natural fibres. Volume 2: Processing and applications**
Edited by R. Kozlowski

120 **Functional textiles for improved performance, protection and health**
Edited by N. Pan and G. Sun

121 **Computer technology for textiles and apparel**
Edited by Jinlian Hu

122 **Advances in military textiles and personal equipment**
Edited by E. Sparks

123 **Specialist yarn and fabric structures**
Edited by R. H. Gong

124 **Handbook of sustainable textile production**
M. I. Tobler-Rohr

125 **Handbook of natural fibres. Volume 1: Types, properties and factors affecting breeding and cultivation**
Edited by R. Kozlowski

126 **Handbook of natural fibres. Volume 2: Processing and applications**
Edited by R. Kozlowski

127 **Industrial cutting of textile materials**
I. Viļumsone

128 **Colour design: theories and applications**
J. Best

129 **False twist textured yarns**
C. Atkinson

130 **Woven textiles: principles, developments and applications**
K. Gandhi

131 **Textiles and fashion: materials design and technology**
R. Sinclair

132 **Modelling, simulation and control of the dyeing process**
R. Shamey and X. Zhao

133 **Advanced process control in textile manufacturing**
A. Majumdar, A, Das, R. Alagirusamy and V. K. Kothari

134 **Functional nanofibres and applications**
Q. Wei

135 **The global textile and clothing industry**
R. Shishoo

136 **Simulation in textile technology: theory and applications**
D. Veit

Preface

As a descendant of a hemp and flax spinning mill owning family; having graduated with a degree in textile science; and having later obtained a PhD in man-made fibre extrusion at Bradford University, it was inevitable that I should pursue an industrial career in textiles. In 1977, my industrial career started in the Synthetic Fibres Laboratory of Courtaulds Plc, Coventry, England, where I was involved in various development activities for the company's POY and LOY polyester spinning operations in Northern Ireland and Eire.

Three years later I joined Ernest Scragg and Sons Ltd (later in 1982, Rieter-Scragg Ltd) as Senior Textile Technologist. 'Scragg' was a leading manufacturer of draw texturing machines for the international market and was renowned for its process know-how and innovations. For a young industrialist, the global world of texturing machine manufacturing provided a wealth of international contacts and excitement. Unknowingly at the time, Scragg was to become highly influential in my industrial career, providing a foundation for both process knowledge and practical experience in draw texturing. In particular, it brought into my life a fascinating textile process, where complex process parameter interactions influence texturing performance and where hands-on experience, together with a systematic approach, are vital for achieving optimum results. Moreover, during the mid-1970s to 1980s research and development activities in texturing were extremely active throughout Europe, both in machine manufacturing and yarn processing. As a consequence, I was fortunate to work alongside many of the leading international figures in research institutions and companies who were associated with draw texturing at the time.

My interests in the draw texturing process grew rapidly during my employment in Scragg and it was not surprising that my career advance evolved around draw textured yarns. To acquire experience in a disciplined yarn production environment I moved to Germany, working for a number of years in polyester and nylon textured yarn process development at Rhodia AG. Yarn end-user experience was also essential, so a period followed in hosiery manufacturing as Technical Director, Aristoc, Courtaulds Textiles. Finally, I returned to the texturing process itself as Technical

Director, Rieter-Scragg Ltd and Managing Director, Intex Yarns Ltd, a textured- and dyed-yarn manufacturer in the UK. Since 2000, I have been operating as an independent consultant in the texturing industry, continuing my international involvement in development activities, process performance improvements and novel yarns.

In recent years there has been a rapid demise of the continuous filament industry in the Western world coupled with a move to lower cost countries, predominantly in the Far East and Asia. Basically, the draw texturing process is currently on a plateau, where a significant advance in process technology is difficult to realise and profit margins are squeezed to the extent that draw textured yarn manufacturing can largely survive only in lower labour cost regions. Moreover, a relocation of the draw texturing machine manufacturing industry has followed yarn processing due to the same cost constraints. The machines, however, tend to be copies of past European designs and significant advance in designs is restricted due to lack of fundamental draw texturing process knowledge. Indeed, the many people that were involved in the pioneering days of the industry are now retired and the number of draw textured yarn- and machinery-manufacturing companies in Europe has drastically shrunk. Moreover, as a consequence, some fundamental key process knowledge and experiences have been lost. It is of course possible to survey past literature on the subject of draw texturing, but this can be time consuming to locate and retrieve and it is often difficult to relate both research and commercially orientated papers to specific problems that can occur in every-day industrial practice.

With my experience in machine manufacturing and process technology in draw texturing exceeding 30 years, together with a unique experience in the synthetic filament supply chain from extrusion through to retailer, I am able to provide an ideal platform for the conveyance of a broad experience and depth of knowledge to existing persons in, and newcomers to, the draw texturing process. Indeed, I very much welcome the opportunity to pass to others my experience and knowledge gained over the enjoyable years in this industry.

This book provides an overview of the properties of, and the applications for, draw textured yarns. It describes key fundamentals in the draw texturing process, the effects of process parameters on texturing performance, and new developments. The key features of draw texturing are initially described for acquiring a basic understanding of the process, followed by details on process optimisation methods and solutions to problems that are commonly encountered in draw texturing. It is a practical guide to the draw texturing process.

Colin Atkinson
E-mail: marcols@supanet.com

1
The development of textured yarns

Abstract: This chapter reviews the development of textured yarns, from artificial silk through to melt spinning. It assesses how fabric properties can be improved by spinning.

Key words: false twist texturing, melt spinning.

1.1 Artificial silk: an early history

In 1710, René A. F. Réaumur,[1] a French naturalist and physicist, recorded in his 'l'Histoire des Insectes' the possibility of making artificial silk: 'Silk is only a liquid gum which has been dried, could we not make silk ourselves with gums and resins?' Robert Hooke had already been inspired in 1664 to consider the possibility of imitating the silkworm to make artificial filaments.[2] It was, however, not until 1855 that George Audemars,[3] inspired by Friederich Schönbein's discovery of nitrocellulose, produced continuous filament from a solution in alcohol and ether: 'Into the mixture thus prepared a steel point is dipped and thus a thread is drawn from the surface of the liquid; this thread is connected with a winding machine, by which it is drawn out until the liquid is exhausted'. Audemars' method can be described as the first evaporative means of producing man-made filaments as it involved solvent evaporation during filament formation. It was never put into use because of the flammability of cellulose nitrate and the high cost of solvents.

'Chardonnet Silk' was the first artificial silk to be manufactured commercially. Chardonnet[4] extruded a hot solution of cellulose nitrate in ether and alcohol through a fine diameter glass tube, and subsequently dried and stretched the filaments in warm air. (Earlier, Joseph Swann[5] in 1883 had spun cellulose nitrate filaments for industrial end-use in electric light bulbs.) The cellulose nitrate industry failed, however, largely due to the price of filaments exceeding that of natural silk and the highly inflammable property of cellulose nitrate.

Cellulose nitrate was indeed not an ideal starting material, and during the infancy of the man-made fibre industry, attempts to spin artificial silk from non-cellulosic materials were not lacking: Gérard,[6] for example, patented a technique for extruding protein fibres into hot air but his process was not commercially exploited. In 1899, 'Vandura Silk',[7] an inexpensive pure gelatine filament, was produced commercially by a similar method, but the product was extremely sensitive to water, despite hardening with formaldehyde vapour, and it disappeared from the market.

During this period, another significant process materialised from the developments of Cross and Bevan.[8] They discovered that if cellulose is treated with caustic soda and subsequently with carbon disulphide, a soluble, viscous compound, cellulose sodium xanthate ('Viscose') is formed. This viscous liquid could be coagulated into filaments by extruding into a bath of ammonium chloride. In 1900, Topham invented centrifugal spinning, where the extruded Viscose filaments were fed inside a rotating container, known as the 'Topham box', to form a cake. This invention overcame many difficulties associated with Viscose processing at the time. Viscose, of course, is still produced today using the coagulation process, which is commonly known as wet spinning.

The discovery of cellulose triacetate in 1869 by Paul Schutzenberger[9] also proved a significant milestone in early synthetic filament development. Although cellulose triacetate was predominantly spun by the wet spinning technique, patents suggest that it could be spun into a hot environment from chloroform or tetrachloroethane,[10,11] whereby the solvents are evaporated, resulting in continuous filaments. This process is known as dry spinning, and it could achieve higher production speeds than the alternative wet spinning process.

In 1904, Miles[12] partially saponified primary acetate (triacetate) into secondary acetate (diacetate) and disclosed the latter's solubility in an easily recoverable and less toxic solvent, acetone. Cellulose diacetate, when dissolved in acetone, forms a viscous solution, later referred to as 'dope', and in World War I it proved to be of major importance for varnishing and water proofing fabric-covered aircraft wings. After the war, a factory that had been erected by Camille and Henri Dreyfus for the manufacture of such dope served no further purpose unless an alternative use for secondary acetate could be found. Attention was turned to artificial fibres, applying the dry spinning method. This proved to be a significant stage in synthetic filament development; acetate filament yarns are still produced today by dry spinning techniques and, because of their uniform quality, dye-ability, versatility, drape and handle, maintain a limited position in women's apparel and home furnishings. A major growth area for acetate filament over the last decades has been for tow, used in the manufacture of cigarette filters.

1.2 The arrival of melt spinning

The start of a new era in synthetic fibres materialised with the discovery of nylon and polyester, which led to the melt spinning process. This proved to be of major importance. Previously, artificial fibres such as cellulose nitrate, cellulose acetate and viscose all had been developed using a natural cellulosic polymer, such as that present in wood or cotton. The breakthrough came in the late 1930s when Dr. Wallace H. Carothers, Du Pont, discovered a means of synthesising a polymer to form the first true synthetic fibre.[13] Carothers discovered that polyamides could be made from diamines and diacids, whereby the two molecules of this condensate would react to give a long molecular chain, suitable for fibre formation.

From these developments, two main forms of polyamides suitable for fibre formation emerged (Fig. 1.1):

- Nylon 66 produced from adipic acid and hexamethylene diamine ('66' because each raw material component contains 6 carbon atoms).
- Nylon 6 made from caprolactam.

In 1941, from the development work carried out by Carothers, two UK chemists J.R. Whinfield and J.T. Dickson developed 'Terylene', a polymeric ester, which was formed by reacting terephthalic acid with ethylene glycol to form polyethylene terephthalate (Fig. 1.2).[14]

This was the second major advance in synthetic fibre development; both nylon and polyester polymers are thermoplastic and, as such, the polymer melts into a viscous mass at high temperatures. By extruding the polymer

Nylon 66 Polymer:

$NH_2(CH_2)_6NH_2$ + $COOH(CH_2)_4COOH$

Hexamethylene diamine — Adipic acid

↓

$NH_2(CH_2)_6NHCO(CH_2)_4COOH + H_2O$

↓

$H(\text{-}NH(CH_2)_6NHCO(CH_2)_4CONH(CH_2)_6NHCO(CH_2)_4CO\text{-})_{\boldsymbol{n}}OH$

Polymer chain repeat unit

Nylon 6 or Polycaprolactam:

$H(\text{-}NH(CH_2)_5CO\text{-})_{\boldsymbol{n}}OH$

Polycaprolactam, polymer chain repeat unit

1.1 Nylon 66 polymer and nylon 6 or polycaprolactam.

$HOOC\text{-}C_6H_4\text{-}COOH$ + $HO\text{-}CH_2\text{-}CH_2\text{-}OH$

Terephthalic acid — Ethylene glycol

↓

$(\text{-}OC\text{-}C_6H_4\text{-}COO\text{-}CH_2\text{-}CH_2\text{-}O\text{-})_{\boldsymbol{n}} + H_2O$

Polyethylene terephthalate, polymer chain repeat unit

1.2 Polyethylene terephthalate polyester (PET).

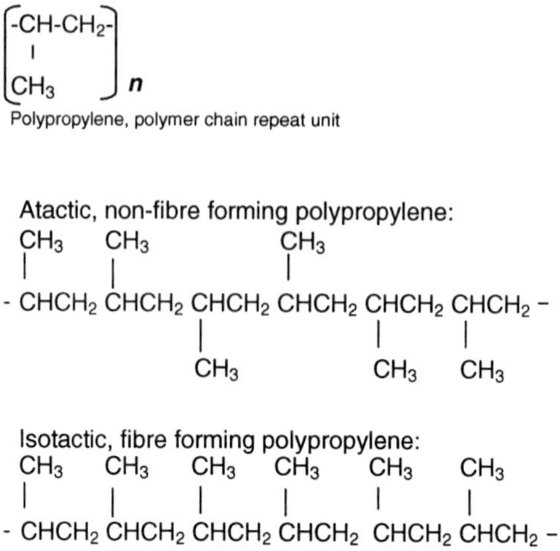

1.3 Polypropylene.

melt through fine holes and cooling, filaments are formed. This is the basis of melt spinning, which is the main production method used for synthetic fibres today.

In addition to the discovery of nylon and polyester, polypropylene emerged in 1954. Polypropylene is a thermoplastic polymer and propylene is a by-product from oil refineries. Karl Rehn and Giulio Natta in Italy were first to polymerise polypropylene.[15] Polypropylene, with its methyl side-chains, however, tends to form an atactic or irregular polymer, which is amorphous and non-crystalline and does not have good mechanical properties suitable for fibre end-use. The development of isotactic polypropylene by Ziegler and Natta followed. They used catalysts to form the methyl groups into a repeating, orderly fashion, situated on the same side of the plane of the main carbon chain to overcome this limitation (Fig. 1.3). As a result, polypropylene, too, became a part of the melt-spinning industry for synthetic fibres. Today, the melt spinning industry for synthetic fibres comprises mainly polyester and nylon; because of its limited thermal and colouration (dye-ability) properties, polypropylene tends to be restricted to specialist markets.

1.3 The melt spinning process

Thermoplastic melt-spun yarns comprise long molecular chains of polymers of high molecular weights. The degrees of orientation at which the molecules lie to the filament axis is influenced by the melt-spinning speed. Most of the filament yarn stretching occurs between the points of extrusion and where the filaments, under tension from the subsequent drawdown feed device, are still in their plasticised state. These molecules are also randomly cross-linked, through chemical- and polar-bonding, with adjacent molecular chains, creating a largely amorphous molecular structure. The terms used to describe the spun yarns refer to spin speeds:

- LOY, low orientated yarn, spin speed around 1200 m/min.
- MOY, medium orientated yarn, spin speed around 2000 m/min.
- POY, partially orientated yarn, spin speed around 3500–5200 m/min.
- HOY, high orientated yarn, spins speed around 6000 m/min.

Little of the early LOY spin speeds are used today, mainly due to the following:

- Originally, within the spinning process, a heated metallic grid system was used to form a polymer melt pool, suitable for extrusion (Fig. 1.4). This early system was prone to polymer decomposition, inhomogeneity and

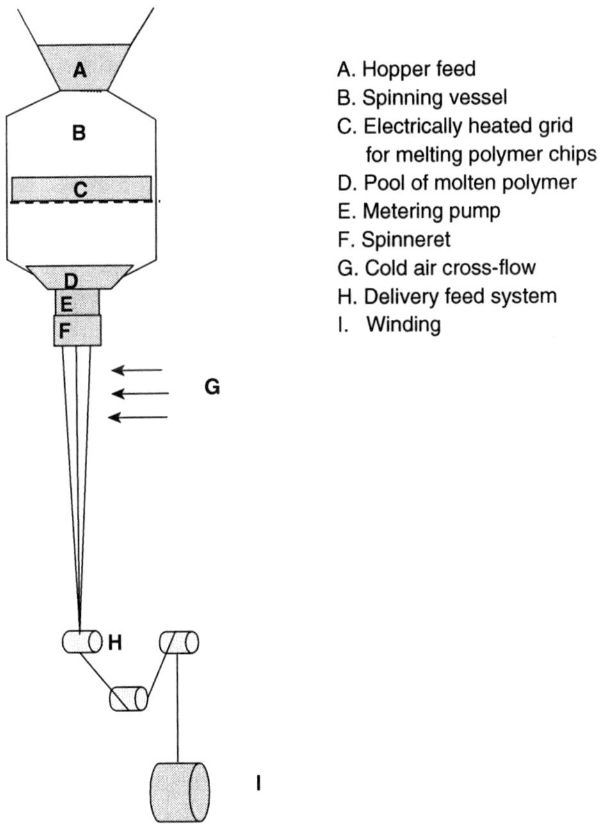

1.4 Early melt spinning process using heated metallic grid to melt polymer.

1.5 Polymer chips.

discolouration, and throughput speeds were limited. The technology was superseded by more modern screw-extruder polymer melt feeds.
- Their low molecular orientation and crystallinity resulted in a poor yarn shelf-life, as molecular structures changed due to stress relaxation and moisture conditions. This adversely influenced dye uptake and colour consistency in fabrics. They had, therefore, to be subjected to a subsequent draw process to stabilise the yarn within a certain time-frame after spinning.
- Higher polymer throughput as a result of process speed advance (POY), due to extruder technology and high-speed winding mechanisms, reduced production costs and provided adequate yarn shelf-life. Moreover, POY could be subjected to simultaneous draw texturing, without the need for any pre-draw process, a point that is discussed in Chapter 2.

Prior to melt spinning, polymer chips (Fig. 1.5) are produced by cutting the solidified strands of extruded polymer from the polymerisation process (although continuous polymerisation–spinning processes are sometimes used). Titanium dioxide is added during polymerisation to achieve desired lustre, when a matt appearance is required. Polymers are usually supplied as bright, semi-dull or dull variants.

In the modern day POY spinning process, the polymer chips are fed, under controlled moisture conditions and in an inert atmosphere, using a screw extruder, which has accurately controlled temperature zones to progressively melt the polymer. Constant melt pressure is maintained by automatic regulation of the extruder speed. From the extruder exit, the melt is fed, under controlled temperature and pressure, to the spinning beam. On exiting the beam, the melt is distributed to a number of spinning gear pumps that accurately feed the polymer to spin pack assemblies, which comprise filtration medium and the spinnerets (Fig. 1.6).

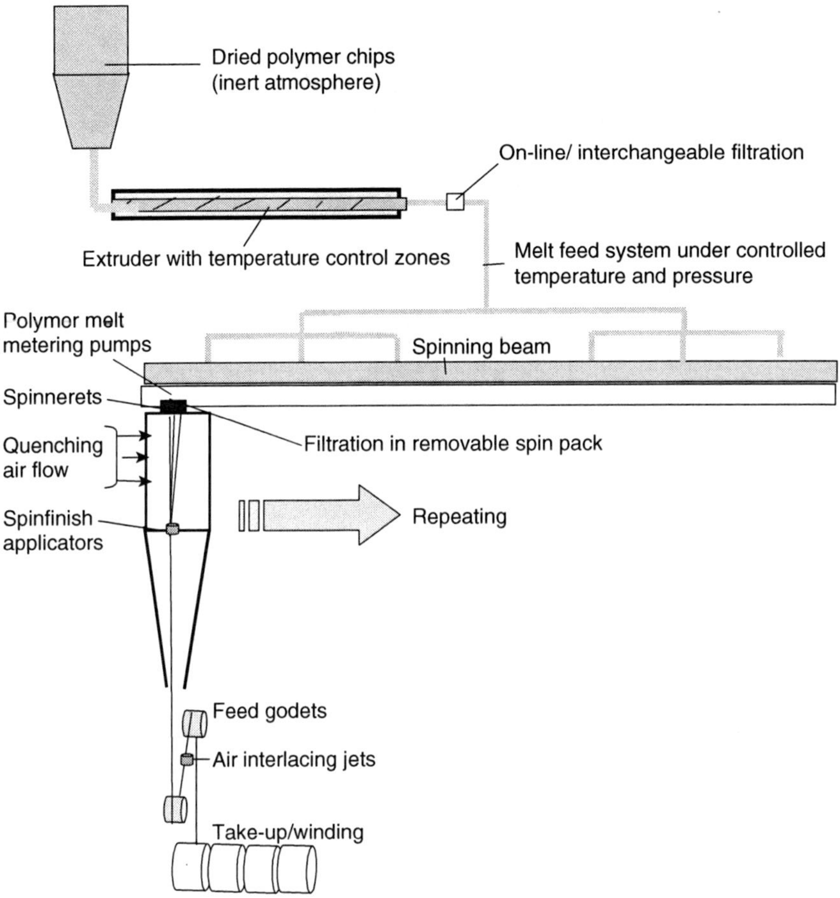

1.6 Schematic diagram of melt spinning process.

Filtration media can typically be particulate, e.g. sand or metal powder, or metallic meshes or webs, and their make-up and composition is critical for desired operating melt pressures to be maintained and for pressure consistency between spinning positions. The spinnerets are stainless steel discs containing precision-engineered holes of around 0.2 mm diameter, through which the filaments are extruded into a cooling airflow from a quenching cabinet (Fig. 1.7). The spinneret holes have specific conical entries and length:diameter ratios for optimum performance regarding rheological flow of the polymer melt. Modified hole cross-sections can be used for making different filament cross-sections, e.g. trilobal or multi-lobal, which affect reflected light and hence the lustre of the yarn. The linear density (count) of the spun yarn is determined not by the diameter of the

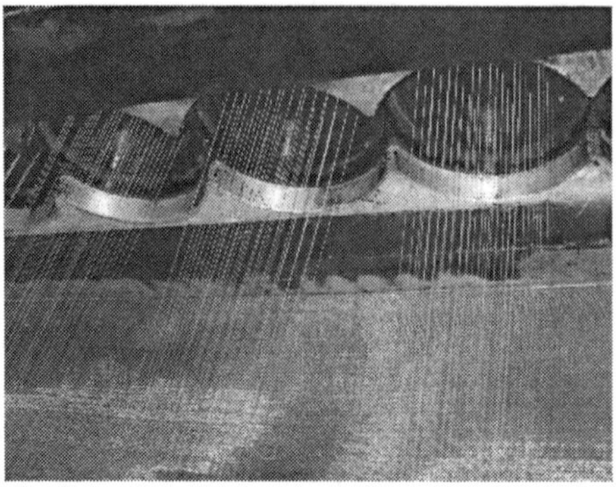

1.7 Formation of a continuous filament yarn; extrusion from a spinneret into the air quenching zone.

1.8 Ceramic applicator for metered spin finish.

spinneret holes but by the throughput of the polymer melt and the speed at which the continuous filament yarn is wound.

The spun yarn is also lubricated by applying a spin finish through metering jets (Fig. 1.8) within the quenching zone. In addition, it is standard practice to apply filament interlacing to the yarn, using compressed air jets before winding to apply an inter-filament cohesion frequency, typically around 4–5 entanglements per metre, to avoid filament separation during off-wind of the spun yarn package in downstream processing (Fig. 1.9).

On drawing, and in some cases also applying heat to the POY, the molecular structure becomes more orientated and there is a large increase in the structured or 'crystalline' regions of the chains (Fig. 1.10). This gives the yarn higher modulus, tenacity and stability, i.e. useable mechanical properties for a textile yarn (Table 1.1).

1.9 Twelve position SPT interlacing jet for spinning processes. (Photo supplied by RPE Technologies GmbH, Germany.[19])

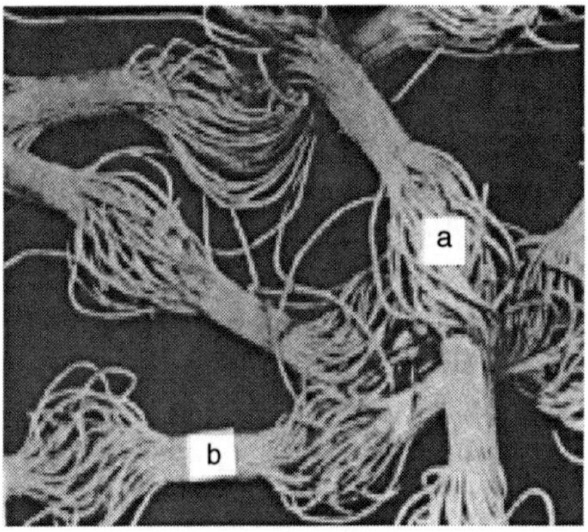

1.10 Model depicting amorphous and crystalline regions in the molecular chains.[16] (a) Amorphous areas (b) Crystalline ordered regions.

Table 1.1 Feed yarn property comparisons; LOY, POY and drawn polyester[17]

	LOY	POY	Drawn
Orientation (birefringence)	0.005	0.038	0.18
Crystallinity (%)	0	5	30
Density (g/cm^3)	1.338	1.341	1.380
Decitex	552	273	167
Tenacity (g/dtex)	1.1	2.0	3.9
Elongation (%)	400	130	30
Modulus (g/dtex)	20	27	99
Shrinkage, boiling water (%)	44	60	8

1.4 Achieving desirable fabric properties through texturing

Fabrics manufactured from filament silk, and indeed unmodified continuous filament synthetic yarns, have characteristics that are very different to yarns spun from staple fibres such as cotton or wool, staple synthetic fibres, or staple fibre blends such as cotton and polyester or wool and nylon. Continuous filament yarns, when simply drawn after spinning to produce desirable mechanical properties, tend to exhibit smoothness, evenness, and parallelism compared to the less regular, more bulky and hairy staple-fibre spun yarns.

Of course, continuous filaments from synthetic fibres can be cut into staple lengths through a separate filament cutting process to form filament staple suitable for a conventional spinning process using raw synthetic- or blended-staple fibres. This is known as 'tow to top' conversion. However, this is a multi-stage process, which is costly, utilising an end-spinning process that was developed for natural fibres. As a consequence, in the 1950s and 1960s, processes referred to as 'texturing' were developed. Texturing imparts desirable textile properties to continuous filament yarns without destroying their continuity through introducing distortions or crimp along their lengths. As a result, textured yarns have suitable volume, stretch and recovery, and air porosity for every day use in fabrics for a wide range of textile end-uses. Initially, the process comprised the insertion of twist into nylon or polyester yarns, thermally setting the twist by steaming in an autoclave, cooling and then untwisting the yarn by the same number of turns as originally inserted. Texturing processes are aimed at capitalising from raw material yarn properties rather than producing a material that fits the traditional yarn process routes.

Thermoplastic, melt-spun continuous filament yarns provide the ideal platform for texturing, as imparted filament distortions can be softened through heat and set by cooling. The texturing process, therefore, evolved around key synthetic thermoplastic filament yarns, which soften when heated and reset when cooled, namely:

- polyester,
- nylon,
- and, to a lesser degree, polypropylene.

Throughout the 1960s and 1970s, various texturing methods were introduced commercially to the yarn manufacturing industry:[18]

- *Edge Crimping* – drawing a thermoplastic yarn over a heated edge, creating differential internal stresses in the filament cross-sections (Fig. 1.11). Edge-crimp yarns exhibit high stretch; they were used for ladies hosiery and circular knit fabrics.

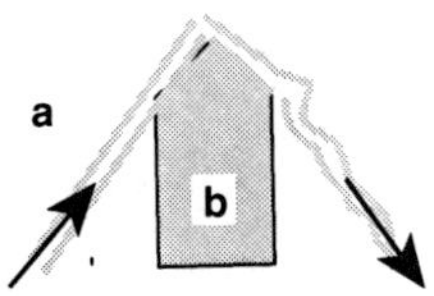

1.11 Edge-crimping principle. (a) Thermoplastic yarn fed over heated knife-edge (b) Knife-edge heater.

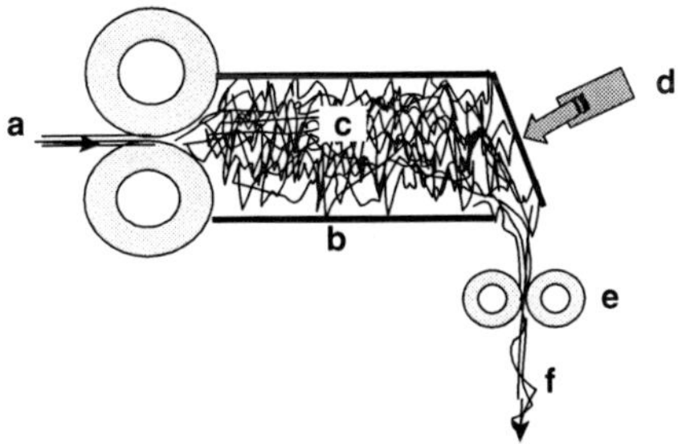

1.12 Stuffer-box texturing principle. (a) Thermoplastic feeder yarn (b) Heated chamber (c) Crimped yarn under pressure from overfeed (d) Pressure control (e) Output roller feeds (f) Crimped, heat-set yarn.

- *Knit-de-Knit* – knitting the thermoplastic yarn on a small diameter circular knitting machine. The plain knit fabric is then heat set and subsequently de-knitted and wound onto a package. After de-knitting, the yarn is deformed according to the knitted loop shapes, forming a three-dimensional structure.
- *Stuffer-box* – thermoplastic yarn is overfed into a heater cylinder, under a pressure from the feed-roller delivery that exceeds that of the controlled outlet resistance (Fig. 1.12). As a result of this process, the filaments are heat-set in a buckled/crimped form before exiting the stuffer-box and being wound onto a package.
- *False Twist Texturing* – inserting high twist levels into thermoplastic yarns, setting the twist by heating and cooling prior to de-twisting.
- *Air Jet Texturing* – air-entangling of continuous filament yarns, applying overfeeds, draw and heat set to the individual component yarns to create filament loops and entanglements.

Of the above methods, the most significant texturing processes to evolve over the last 50 years are *False Twist Texturing* and *Air Jet Texturing. False Twist Textured Yarns* (Fig. 1.13) comprise a multitude of crimp in the individual filaments and are inherently elastic. The degree of crimp and hence elasticity can be determined largely by the amount of twist applied to the

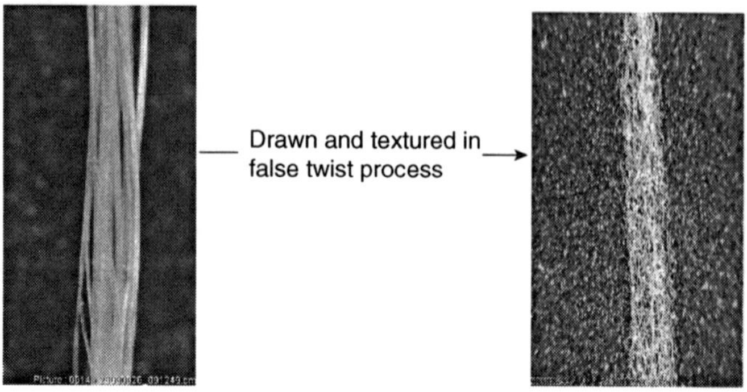

1.13 POY and false twist textured nylon 66 yarns.

1.14 Scanning electron micrograph of air jet textured yarn.[20]

yarn in the heat-set process. The modern process for false twist texturing is based on a technique patented by Finlayson and Happey in 1933, whereby temporary twist is imparted to a moving yarn upstream of a spindle (twist applicator). The twist is thermally set and cooled before becoming automatically untwisted after the spindle.[21]

Air Jet Textured Yarns (Fig. 1.14) provide a staple fibre appearance due to the multitude of filament distortions and entanglements brought about by a high-pressure air flow as they pass through a purposely designed air jet.[20] They exhibit low elasticity due to their filament construction and because of their filament entanglements, and they have good abrasive characteristics.

Regarding *Edge Crimping, Knit-de-Knit and Stuffer-box Texturing*, only the *Stuffer-box* technique has continued to be developed; it is used for high decitex carpet yarns. Old *Knit-de-Knit* machinery is still, however, used today in very limited applications for textured yarn specialities and where fashion demands a low crimp/low elasticity appearance.

1.5 References

1 C. de Cizancourt, *Ciba Review*, 2, 3, (1967).
2 H. R. Mauersberger, E. W. K. Schwarz, *Rayon and Staple Fibre Handbook*, Barnes Printing Co. Inc., New York, p. 1, (1939).
3 G. Audemars, BP 283 (1855).
4 H. Chardonnet, DP 56,331 (1891).
5 J. Swann, BP 5978 (1883).
6 M. P. E. Gérard, DP 40,373 (1886).
7 A. Millar, *J. Soc. Chem. Ind.*, 31 Jan., 16 (1899).
8 C. F. Cross, E. J. Bevan, *Cellulose: an Outline of the Chemistry of the Structural Elements of Plants*, http: books.google.co.uk
9 M. W. Alford, *Journal of the Textile Institute*, 52, 243, (1961).
10 Friedrich Bayer and Co., BP 28,733 (1905).
11 L. Lederer, BP 6751 (1905).
12 G. W. Miles, USP 838,350 (1906).
13 R. R. Hedge, A. Dahiya, M. G. Kamuth, *Nylon Fibers*, www.engr.utk.edu/mse/Textiles/Nylon%20fibers.htm
14 R. R. Hedge, A. Dahiya, M. G. Kamuth, *Polyester Fibers*, www.engr.ukt.edu/mse/Textiles/Polyester%20fiber.htm
15 www.articlesbase.com/online-business-articles/polypropylene-2851412
16 H. F. Mark, S. M. Atlas, E. Cernia, *Man-made Fibers: Science & Technology*, 1, 305, Interscience Publishers, John Wiley & Sons Inc., New York (1967).
17 O. L. Shealy, W. A. Lanka, R. E. K., Draw Textured Feed Yarns, *Modern Textiles*, July, 56 (1975).
18 I. B. Piller, *Bulked Yarns Production, Processing and Applications*, SNTL Publishers of Technical Literature, Prague, Czech (1973).
19 *Interlacing Jets for the Synthetic Fibre Industry*, Technical Literature, RPE Technologies GmbH, Germany.
20 D. K. Wilson, The Production of Textured Yarns by Methods other than the False Twist Technique, *Textile Progress*, 16, 3 (1987).
21 M. J. Denton, *The Throwsters 'from Silk to Synthetics'*, British Throwsters Association (1994).

2
Basic principles of false twist texturing

Abstract: This chapter reviews the basic principles of false twist texturing. It also introduces twist application methods.

Key words: false twist texturing, application methods.

2.1 The concept of false twist texturing

The primary function of the false twist texturing process takes advantage of the thermoplastic nature of polymers. Basically, continuous filament feeder yarns are:

- *Drawn* (to bring about molecular orientation and crystallinity).
- *Twisted* (individual filament deformation within the yarn).
- *Heat set* in their twisted form (molecular crystallinity and orientation from heat energy and yarn tension).
- *De-twisted* (under temperature conditions and tension that do not affect the heat-set filament deformations).

In the early days, spun LOY yarn tended to be pre-drawn on draw twisting machines after the spinning operation, to provide desirable, stable yarn mechanical properties for subsequent use in false twist texturing. This intermediate process, together with the small size of the yarn packages (1.5–2.0 kg) on kops, proved costly. As a result, two single-stage texturing processes evolved after the spinning operation:

- *Sequential* draw texturing, and
- *Simultaneous* draw texturing.

In the *sequential* method, the feeder yarn is pre-drawn within the process to attain the desired yarn tensile properties (tenacity and elongation). The older LOY materials in general had to be textured by this method because threading a low orientated/low crystalline yarn on the texturing heater is very difficult via the alternative simultaneous method. Pre-drawing provides both molecular orientation and crystallinity to support trouble-free threading via the sequential route.

In *simultaneous* draw texturing, the feeder yarn is simultaneously drawn, twisted and heat-set prior to de-twisting in a continuous process. Since the introduction of POY spinning in the early 1970s, the simultaneous method has become the norm in false twist texturing. It offers the following advantages over the sequential method:

- Better ergonomics (more operator friendly in yarn threading on the machine).
- Higher attainable yarn bulk (higher energy is required to heat-set pre-orientated, crystalline yarn).
- Lower energy consumption (no extra feeds and reduced heater set temperature).
- Better process efficiencies (tendency for broken filaments in the sequential draw zone can create yarn breaks in the subsequent texturing zone).

Sequential draw-texturing still exists today, where very limited quantities of LOY are still spun with nylon 66 in vertical operations. Sequential zones, incorporating extra yarn input feeds in draw texturing, are also re-emerging for the processing of fancy yarns (see Chapter 9, Draw textured yarn variants and speciality yarns).

The functions of the sub-sections of the primary texturing zone and the basic yarn parameter profiles in generating the required structural properties of the yarn for achieving tensile performance, elasticity and bulk, are outlined in Fig. 2.1 for both sequential and simultaneous processes. For sequential false twist texturing, the yarn is drawn in a pre-draw zone prior to the twist-heat set operation. For polyester use, uniform draw to the yarn is achieved with the use of hot pins, where the yarn is heated to temperatures around 80–100°C and drawn at the exit of a heated stationary cylinder (diameter usually 60 mm or 80 mm). Nylon can be cold drawn and small diameter ceramic pins are used for this purpose (Fig. 2.2).

The twisting torque in the yarn immediately before its entry into the heater is important and this is often overlooked. It is here that filament migration begins. The filaments try to migrate into a simple helical form, but in practice the migration is limited due to frictional and bending constraints. As a result, they migrate in and out of the twisted filament bundle. The migration characteristics are influenced by:

- Inter-filament friction (spin finish type and level applied to the feeder yarn, filament cross-section and lustre).
- Twisting torque (twist level applied, friction and angles of wrap within the texturing zone upstream of the twist applicator).

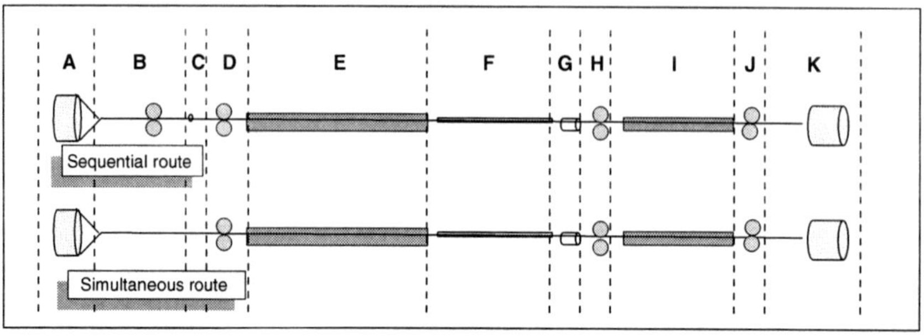

Zone	Description	Simultaneous process	Sequential process
A	Creel supporting feeder yarns	Due to 'over end' package off-wind, low twist levels are imparted, increasing with package diameter reduction. This is not a problem in the process, although it can affect special high-torque textured yarns (see Chapter 4, Section 4.6).	
B-C	Pre-draw	Not applicable	Feeder yarn is drawn over ceramic pin (nylon) or hot-pin (polyester) to create desired yarn tenacity and elongation through molecular orientation and crystallinity. Yarn tensions can increase several fold due to draw force at the point of draw. The ratio of input feed to pre-draw roller speed defines the draw ratio applied to the yarn.
D	Input feed	The ratio of the intermediate feed to the input yarn feed speed defines the draw ratio applied to the yarn. Process tension is strongly influenced by the draw ratio.	The ratio of intermediate feed to input feed speed determines the desired yarn process tension only.
E	Yarn heating	Twisted yarn is simultaneously drawn and heat set. The draw point is located in order of 20-30% along heater length.	Twisted yarn is heat set, no draw point.
		Heat setting of the migrated filaments, together with yarn tension, provides stable mechanical and physical textured yarn properties.	
F	Yarn cooling	The twisted yarn temperature is ideally reduced to less than 90 °C for polyester so that it exits the twist applicator below the glass transition temperature of the polymer. Yarn tension and twist can be above the levels on the heater thread path due to yarn-to-surface contact friction, which is strongly influenced by thread path geometry.	
G	Twisting and de-twisting	Twist is imparted into the yarn upstream of the twist applicator. De-twisting takes place on the exit to the twist applicator. De-twisting yarn temperatures above the polymer glass transition temperature for polyester can adversely affect morphology and hence dye uptake consistency of the process. For nylon yarns the de-twisting tempearature is not as critical. Together with heater set temperature and yarn tension, the twist level upstream of the twist applicator influences the crimp characteristics of the textured yarn, i.e. fabric cover, elasticity and softness.	
H	Output/ intermediate feed	This speed is used to define the speed of the texturing process. It is the reference used for all speed ratios of yarn feed systems in the process.	
I	Secondary heater	Reduces boiling water shrinkage, residual torque and elasticity (mainly for polyester applications).	
J	3rd feed	Determines yarn tension in secondary heater by relative speed to intermediate feed.	
J-K	Lubrication and winding	Lubricant is normally applied to the textured yarn before winding, to facilitate downstream processing.	

2.1 Functions of the sub-sections within false twist texturing.

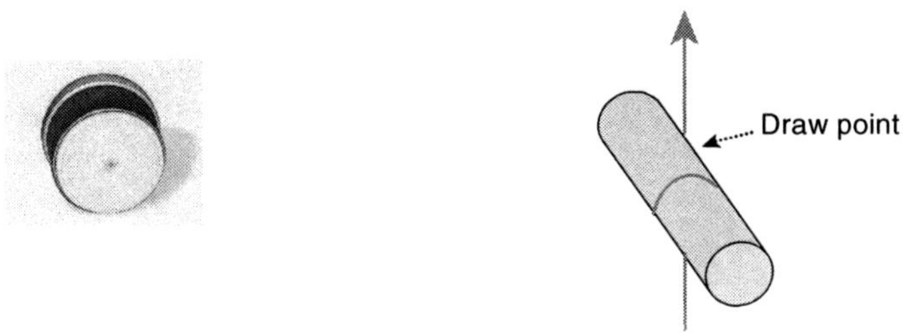

2.2 Hot pin and ceramic draw pin used for drawing continuous filament yarns in sequential false twist texturing (Dienes Apparatebau GmbH, Germany).

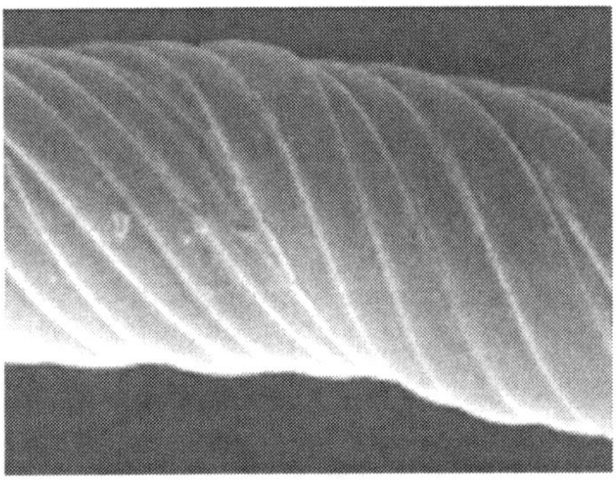

2.3 Twisted, drawn multi-filament yarn on heater exit.

- Number of filaments/filament fineness.
- Yarn tension (draw ratio).
- Process speed.

Heat is usually transferred to the twisted filaments through contact with heated tracks, which are heated to a uniform set temperature by a vapour phase medium. There are, however, alternative heating methods used (see Chapter 3, Sections 3.3–3.5). On exiting the heater, the yarn is in a drawn, twisted state and at high temperature (Fig. 2.3). Downstream of the heater, the yarn is cooled as it passes over a metallic track prior to entering the twist unit. Various cooling track designs are available; these are discussed in Chapter 3, Section 3.7.

For a given draw ratio, increasing applied twist levels within the texturing zone reduces yarn tension (*T1*) immediately before the twist applicator (referred to in the industry as twist unit or aggregate). This is a result of

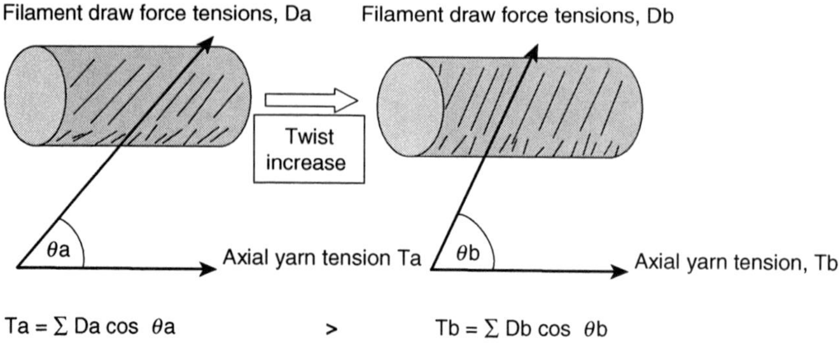

2.4 Reduced yarn tension with increase in twist level.

increased twist helix angle in the twisted yarn that is in a plasticised state, whereby:

- Twist contraction forces are not significant because the yarn is in a plasticized state.
- Axial yarn tension is largely a function of the sum of the individual filament draw forces/tensions.
- The overall filament twist helix angles of the filaments increase with twist increase.

This phenomenon is depicted in Fig. 2.4.

In Fig. 2.1, the schematic diagram shows a thread path that includes a secondary heater. The function of this heater is to reset the crimped yarn in an untwisted and relaxed state at minimum tension. This reduces the high contractive power or stretch in the textured yarn that is inherent from the primary texturing zone. In general, high yarn elasticity is not required in weaving end-use, which predominantly sources textured polyester yarns. Nylon false twist textured yarns have higher elasticity than polyester. They are, therefore, more commonly used in the circular knitting and fine gauge hosiery industries, where high elasticity is desirable. As a result, false twist texturing machines are largely twin-heater and single-heater types for the polyester and nylon industries respectively.

When using a secondary heater, the textured yarn relaxes under heat, releasing internal stresses in the molecular structure of the filaments.[1] For this, high yarn delivery overfeeds are necessary in the heater zone, typically 8% to 20%, for low yarn tension. The heaters tend to be tubular, non-contact to avoid surface-to-yarn friction. Increasing the heater set temperature and, to a lesser degree, decreasing over-feed, reduces the textured yarn crimp and elasticity.

Releasing internal filament stresses through the use of a secondary heater also reduces undesirable twist liveliness in the yarn (Fig. 2.5). Too high a twist liveliness can be problematic due to yarn snarling tendencies in stop–start downstream processes, such as warp application for Raschel knitting.

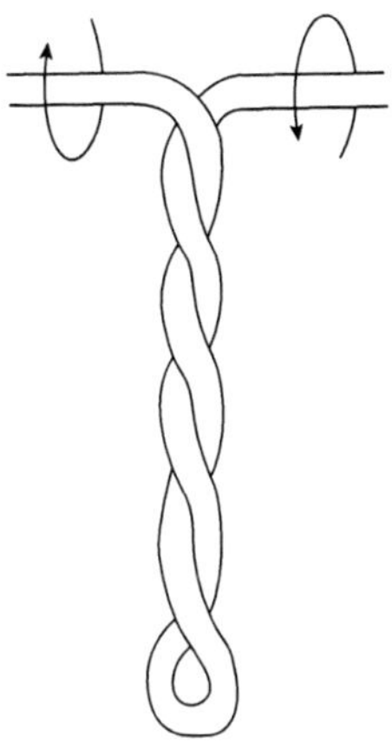

2.5 Self-twist tendency caused by twist liveliness in the textured yarn.

Twist liveliness is residual twisting torque in the yarn, which is inherent from the twist-heat set process in the primary texturing zone. Such residual twist tends to decrease with increase in applied twist level, due to the inherent increase in filament entanglements restricting the filaments to self-twist in a simple helical form. As a result, low stretch cannot be realised by simply applying low temperature and twist in the primary texturing zone, as residual torque levels may be too high and problematic in downstream processing. There are also other constraints regarding yarn quality consistency issues at low twist levels and heater temperatures.

False twist texturing is a multi-position process, comprising typically 216 or 240 positions per machine. For textured yarn quality consistency, it is paramount that:

- Heater track temperatures are maintained within tight tolerances across the machine.
- Yarns feed component speeds between positions are consistent.
- Yarn path geometries are identical across the machine for individual positions (particularly in the primary texturing zone).
- Yarn contact surfaces are consistent across the machine (with respect to ceramic guides, any component wear and cleanliness).

Demands for plant operational disciplines are high for consistency in textured yarn quality to be realised.

2.2 The false twist mechanism

In false twist texturing, the function of the twist application is to:

- Apply twist to the continuous filament yarn upstream of the twist applicator (referred to as twist unit or aggregate) and release the twist downstream of the applicator.

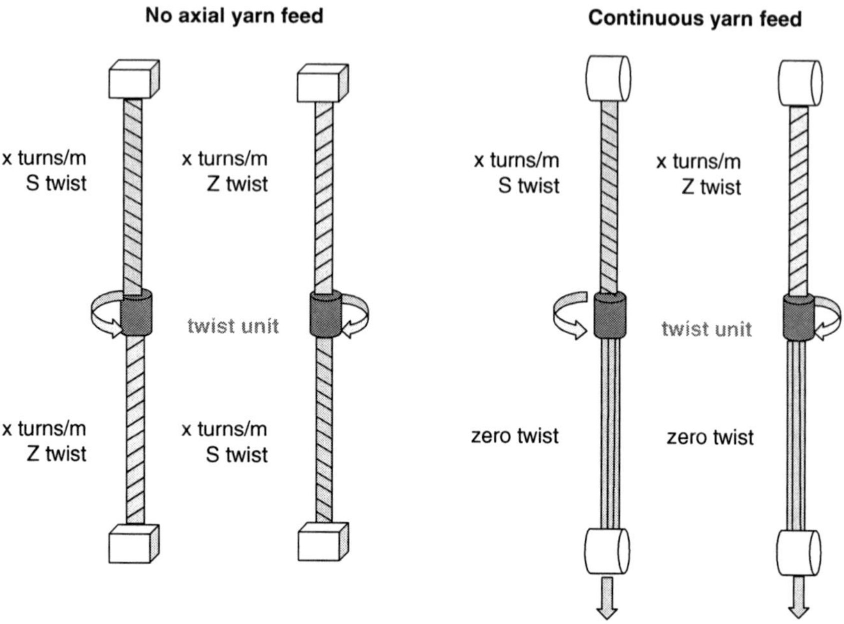

2.6 Formation and release of twist on static and running yarn.

- Apply twisting torque to the yarn upstream of the twist unit so that sufficient twist levels are realised during yarn passage over the heater for desirable filament crimp.
- Provide adequate flexibility for required twist levels to be realised over a realistic range of yarn counts (typically 8dtex nylon to 330 dtex polyester).
- Provide twist level consistency between processing positions on the texturing machine, so as to ensure crimp value and dye uptake consistencies within the textured product.
- Apply twist in a manner that no damage to the filaments is incurred from the twisting operation itself.

The mechanism by which twist is imparted to a running, continuous filament yarn and then continuously released on exiting the twist unit is demonstrated in Fig. 2.6. Basically, referring to Fig. 2.6:

- For a static situation, with no axial movement through the twist unit, an equal and opposite twist to the upstream yarn is formed downstream of the unit.
- It is evident that this applies to both S and Z twist directions.
- When the yarn is fed in an axial direction through the twist unit, the twist imparted to the upstream yarn is cancelled by an equal and

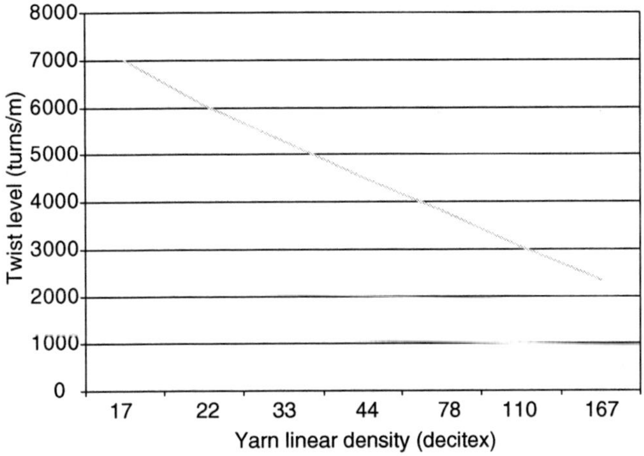

2.7 Typical twist levels experienced in false twist texturing.

opposite twist downstream of the unit. As a result, the yarn has no remaining twist downstream of the twist unit.

This is the basis for the twisting technique applied to false twist texturing, i.e.

- The running yarn is heat-set upstream of the twist unit in its twisted state and is cooled to an operable temperature prior to reaching the twist unit.
- On exiting the twist unit, the filaments immediately de-twist, but retain the crimp from their heat-set twisted form.

Twist levels experienced in false twist texturing vary, largely according to yarn count. Figure 2.7 shows a typical trend in twist level with increase in textured yarn linear density (decitex). This data has been accumulated by taking twisted yarn samples from the primary texturing zone and measuring their twist levels by de-twisting the yarn on laboratory twist measuring equipment. Twisting rates are indeed high and for high-speed low decitex applications can exceed 100 000 turns/sec for current texturing speeds!

2.3 Twist application methods

The most positive twist insertion method used in false twist texturing is the magnetic spindle method (commonly known as 'pin texturing'). Although various twist insertion methods had been developed at the time,[4] pin texturing emerged as virtually the sole method used in this process up to the early 1970s. In very few cases, the method is still used today for twist applications where crimp characteristics, yarn quality consistency and desired fabric appearances are difficult to achieve by alternative methods, i.e. very high

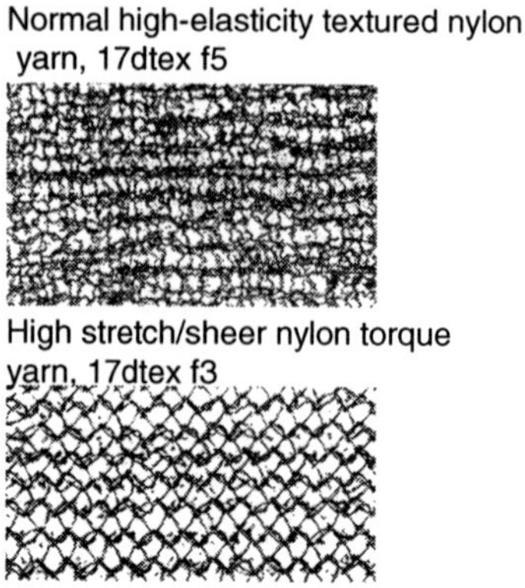

2.8 Sheer effect from torque yarn in 4-feed circular knit for ladies' hosiery.[2]

and very low crimp values that are outside the operating window for more modern, alternative methods of imparting twist. A good example of this is for the nylon torque yarns, typically 17dtex f3, that were used for ladies' hosiery applications in the 1990s. Here, very low twist levels of around 1900 t/m are required to:

- Create an extremely low fabric cover on the leg and high sheer appearance (Fig. 2.8).
- Provide sufficient elasticity to take the form of the leg through high tendency to self- twist via residual torque (Fig. 2.8).

For false twist yarns, the elasticity is brought about by the multitude of filament crimps. This is not the case for torque yarn, where elasticity stems from the tendency of the yarn to self-twist.[2]

The magnetic spindle comprises a rotating sapphire pin around which the yarn is looped. The housing for the pin is fitted with a fine tube at entry and exit, through which the yarn passes. The external surfaces of these tubes are driven by frictional contact on rotating drive surfaces. They are held against the drive surfaces by a magnetic force (Fig. 2.9). The exact twist required in the yarn is determined by;

$$\text{Yarn twist (tpm)} = \frac{\text{rotational speed of spindle (rpm)}}{\text{axial speed of the yarn (m/min)}} \qquad [2.1]$$

The advantages of twisting with the magnetic spindle are:

- Twist flexibility (this can be set through selection of spindle rotation and yarn speeds).
- Twist precision between processing positions.
- High twist levels achievable (high yarn elasticity, fabric softness).
- Low twist levels achievable (sheer effect in fabric).

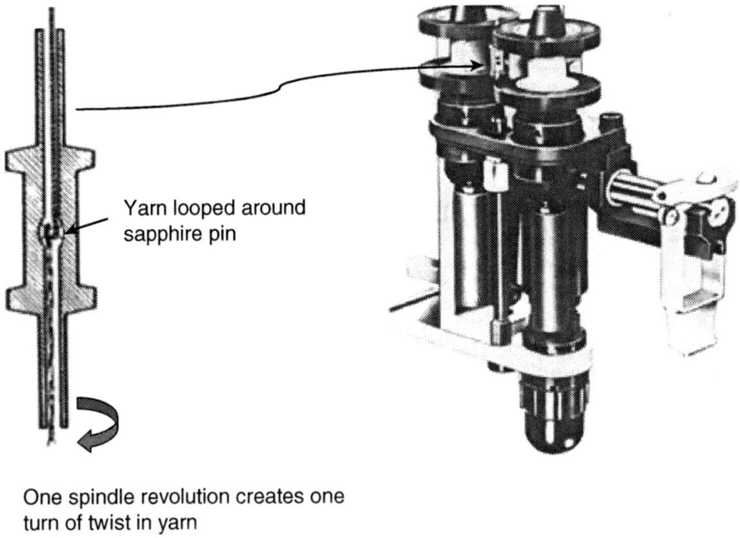

2.9 Magnetic spindle (FAG type MFD 800 magnet spindle unit).[3]

The disadvantages of magnetic spindles are:

- Maximum texturing speeds very limited (due to limited speed of the spindle).
- High noise emission.
- High yarn tension after the spindle (due to frictional forces as the yarn is pulled in an axial direction over the sapphire pin). This can cause broken filaments, which can adversely affect downstream processes. Typical pre- to post-spindle yarn tension ratios are > 2.0.
- Threading of the spindles is slow and labour intensive (using nylon monofilament for threading, ca. 0.3 mm diameter).

Improvements have been made through the introduction of single-motor drives on the magnet spindle units as an alternative to the earlier tangential machine-length drive belts, in order to reduce noise emissions. Mechanical speed limits still, however, impose maximum spindle speeds of up to 800 000 rpm, thus severely restricting the maximum texturing speed attainable and adversely affecting process economics. In the last 30 years, rapid evolution of the significantly more economical friction disc twisting texturing technology has brought about a sudden demise of the magnetic spindle; the latter twisting method is now used only on old machinery and for special yarn requirements.

The most commonly used method of twist insertion in false twist texturing is by friction discs. This technology has superseded the magnetic spindle. Here, the yarn is twisted through frictional contact with rotating friction

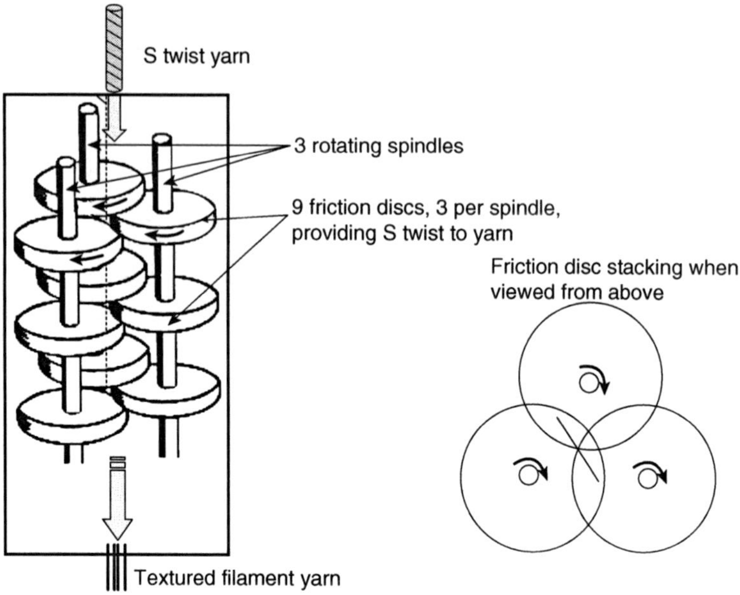

2.10 Yarn passage through friction disc unit.

discs, which are located on a 3-spindle unit. The design of the assembly is such that the yarn passes in a helical path along the central axis of the spindle (Fig. 2.10). Twisting torque is applied to the yarn through friction contact with each rotating disc. As the yarn continuously passes through the friction disc, twist is inserted and remains during passage through the disc stacking (Fig. 2.11). The twist is immediately released to zero twist on exiting the twist unit in accordance with the false twist concept.

The geometries of the friction disc separation and overlap are such that the frictional forces from the disc surface-to-yarn contact generate both twisting and forwarding vectors to the yarn (Fig. 2.12). It is necessary in friction disc system design for the yarn to lie at a pre-determined wrap angle on the discs to provide a forwarding velocity vector. It can be seen, for example, that if the yarn path is simply at right angles to the rotating disc there is no forwarding vector; it would then act in a negative mode, creating high frictional forces opposing the yarn axial direction and high resultant yarn tension (*T2*) downstream of the twist unit. This is a totally impracticable situation, where the yarn is subjected to too high an axial load, and is damaged as a result of high tension and reaction forces on contact surfaces.

Twisting torque is applied to the yarn through contact with each friction disc surface, with the effect that the higher the number of friction discs fitted to the unit, the higher the twist level inserted to the yarn. The disc-to-yarn

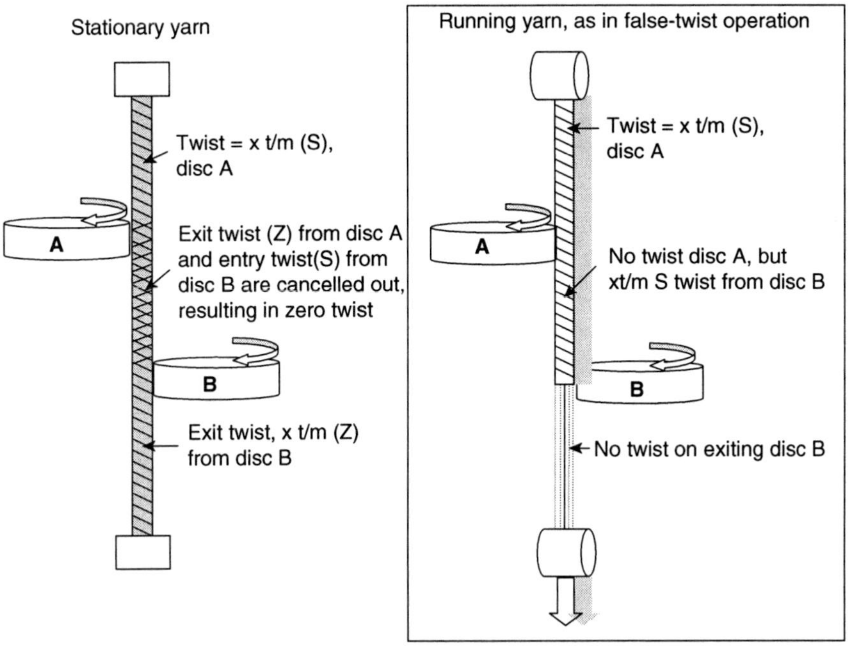

2.11 Twist remains in yarn whilst passing through friction discs and is released on exiting the disc stack.

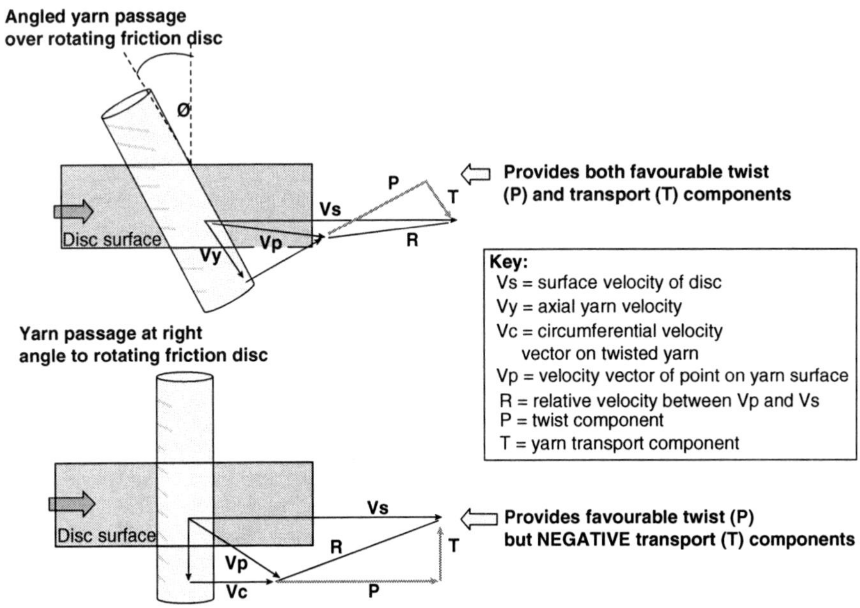

2.12 A yarn angle on the friction disc results in both twisting and forwarding velocity components.

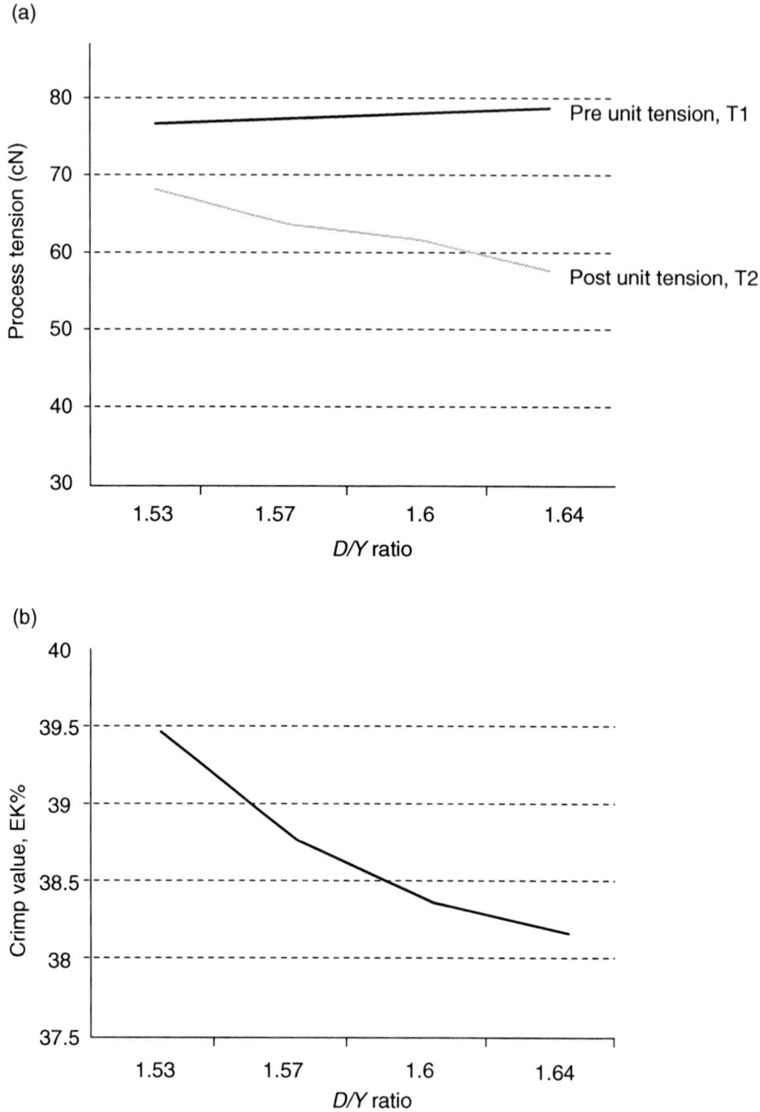

2.13 Example where increase in friction disc speed for fixed process speed and texturing zone conditions has not increased twist level (Yarn type: polyester 167dtex f48; Process speed: 900 m/min).
(a) Influence *D/Y* on process tensions before and after twist unit.
(b) Influence *D/Y* on crimp value, EK%.

speed ratio (*D/Y*) has surprisingly little effect on twist level! Increasing the disc speed tends to increase angle θ (Fig. 2.12), thus increasing the yarn forwarding velocity vector, *T* and as a result reducing the tension in the post-twist unit (*T2*). The pre-twist unit tension (*T1*), which reflects the amount of twist, in general tends to remain steady with increased *D/Y* (It would reduce with increased twist level! – see Fig. 2.4). In some instances, it can even increase as the disc system itself acts as a feed system forwarding the yarn. An example of this is given in Fig. 2.13 where *T1* increases and crimp value (EK%) decreases, reflecting a trend to twist reduction with increased *D/Y*.

For a defined yarn geometry in the friction unit (i.e. spindle separation, disc diameter, material and thickness), yarn twist levels are changed predominantly by the selection of friction disc type and number of discs. The influence of friction twist spindle geometries, disc materials and their limitations, in particular to maximum attainable process speed and textured yarn quality, are discussed in more detail in Chapter 4, Sections 4.3 and 4.4.

2.4 References

1 Sh. Mischinev, Dr. Betschev *et al.*, Untersuchung des Set-Prozesses bei der Falschdrahttexturierung von Polyester-Filamentgarnen, *Textil Praxis International*, 234–237, March (1988).

2 C. Atkinson, A. Babakhani, Texturing of Yarn for Hosiery End-Uses, *Knitting International*, 40–44, Oct. (1991).

3 Monsanto, *Draw-textured Yarn Technology* (1974).

4 J.W.S Hearle, L. Hollick, D.K. Wilson, *Yarn Texturing Technology*, Woodhead Publishing Ltd (2001).

3
Yarn texturing machine design

Abstract: This chapter reviews yarn texturing machine design. It discusses vapour phase, high-temperature and intermediate-temperature heaters, yarn cooling and secondary heaters. It also covers different machine profiles, including folded thread and straight thread path profiles as well as hybrid texturing zone profiles.

Key words: false twist texturing, vapour phase heaters, high-temperature heaters, intermediate-temperature heaters, folded thread path profile, straight thread path profile, hybrid texturing zone profiles.

3.1 The evolution of yarn texturing machines

Over the last 40 years, the textured yarn industry has experienced dramatic process speed increases for both polyester and nylon textured yarns. From a multi-step process up to 1950 and a continuous process in 1955, process speeds of up to 200 m/min materialised in the early 1970s with the industry now experiencing typical commercial production speeds of:

- Polyester: 900 m/min (up to 1000 m/min for 78dtex polyester).
- Nylon: 1100 m/min (fine denier nylon).

In general, the advance is largely attributable to:

- The continuous strive to improve process economics through reduced yarn conversion costs.
- Texturing thread path (heat–cool zone) development through progressive improvement in process know-how.
- The introduction of friction disc twisting and its continued development.
- Continuous improvement in machine mechanical design (drive systems, vibration, reliability of components, reduced noise emission).
- Higher speed wind capabilities of textured yarn packages.
- Development of process monitoring facilities.
- Continuous improvements in feedstock yarn qualities.

By 1970, the machines began to look something like those used today and from here the marked process speed advance was mainly attributable to five generations of machine developments (Table 3.1). In the 1970s, machine

Table 3.1 Five generations of process speed advances in draw texturing

Generation	Year	Maximum process speed (m/min)	Comments
1	<1970[3,36]	approx. 200	– magnetic spindle speed limitations
2	–1975	approx. 450	– introduction of friction disc technology – improved machine mechanical design – simultaneous texturing (POY feeder yarns)
3	–1978[4]	approx. 600 (800, fine yarn counts)	– improved friction disc technology – new machine heat–cool geometry – vapor phase heaters – improvements in POY quality
4	–1985[5,6,40]	approx. 800 (900, fine yarn counts)	– improved friction disc technology – belt twisting technology – improvements in POY quality – online quality control (tension monitoring) – improved package build systems – machine mechanical performance
5	–2009[7]	approx. 900 (1200, fine yarn counts)	– high temperature heaters – shorter heat–cool zones – new machine profiles – improved mechanical drives and controls – lower noise emissions – improved air-intermingling jets for high speed use – continuous improvement in POY quality

manufacturers were largely European and Japanese due to the predominance of the industry in Europe, USA and Japan.[1] Indeed, there were many machine manufacturers in this growth industry at the time:[2]

- Europe: ARCT (France), Barmag (Germany), Berliner Maschinenbau GmbH (Germany), Ernest Scragg and Sons (UK), Davide Giudici and Figli (Italy), Heberlein (Switzerland), Platt (UK), Invista (Czechoslovakia).
- USA: Leesona Corporation.
- Japan: Mitsubishi Heavy Industries Inc., Nihon Spindle Manufacturing Co.

Of these manufacturers, Barmag and Ernest Scragg and Sons Ltd (later Rieter-Scragg Ltd) emerged as the leading manufacturers in false twist texturing technology. Today, however, apart from Barmag and Giudici, none of these machine manufacturers exists, largely due to the relocation of the false twist industry to Asia and the Far East, and the relatively high cost of machine manufacturing within Europe. Currently, the industry is largely supplied by machines manufactured in China and India, which at the moment tend to produce low-cost machines based on the technology developed by the European and Japanese manufacturers in the 1980s and 1990s.

Although mechanical speed capabilities of texturing machine manufacturers are claimed to be up to 1500 m/min speed, in reality these speeds are not yet realised in yarn production plants due to:

- Process speed limitations within the constraints of the physical and mechanical yarn properties required for the downstream processes.
- Machine reliability in terms of component wear, particularly in relation to the package build systems, leading to increased maintenance costs.
- Increased yarn breaks, reduction in production efficiencies and reduced first quality textured yarn quantities adversely affecting the process economics.

3.2 Yarn heating and residence time in the primary texturing zone

The design and construction of heaters has been a subject of much scientific investigation over many years and reflects the importance of the efficiency of the heat setting process. Essentially, the heater has two functions to perform:

- Raising the temperature of the yarn to required levels.
- Maintaining the yarn temperature to a required and consistent level for a sufficient time to allow the restructuring of the inter-molecular bonds.

Fully drawn polyester yarn heats up faster than nylon 66, because of its lower specific heat:

- Polyester, approx. 1200 J/(K.kg)
- Nylon 66, approx. 1700 J/(K.kg)

It also responds more rapidly than nylon to heat-setting[8]; and POY polyester yarn processed by simultaneous draw-texturing responds even more rapidly to heat than in the older sequential method. This is believed to be because:

- the work of drawing and heat of crystallisation raises the temperature of the yarn in the simultaneous process.
- The undrawn POY yarn does not have the stable structure of a fully drawn yarn. In the latter, the molecular structure has to be broken down before the twisted state can be heat-set into the filaments.

Experimental work carried out on PET 167dtex in the late 1970s at draw-texturing process speeds between 150 m/min and 750 m/min demonstrated that in simultaneous draw-texturing, the yarn reached maximum crimp contraction (yarn bulk) values in 0.2 sec residence time on the heater.[8] In texturing with a fully drawn yarn as the raw material, it took 0.4 sec to reach maximum crimp retraction. For both process routes, virtually no crimp was imparted in the yarns up to residence times of 0.07 sec. In addition, at a heater set temperature of 210°C, higher crimp contraction levels were realised via the simultaneous texturing method with POY feeder yarns. The work also highlighted that high processing speeds, especially at low heater set temperatures, have a tendency to reduce the crimp stability of PET yarns; for consistency in crimp stability on a conventional vapour phase heater, yarn residence time on the heater should be not less than 0.2 sec. Similar work conducted on a 78dtex nylon 66 yarn also demonstrated that yarn residence time on the heater must be greater than 0.2 sec for steady-state crimp retraction to be realised.[9] As in the case of polyester texturing, higher crimp retraction levels were reported to be reached by the simultaneous texturing route.

3.3 Vapour phase heaters

In the early days, draw-texturing machines used simple electrical heaters in their primary texturing zones for heating the twisted yarn. These were typically 1.0 m in length for process speeds that did not exceed 200 m/min at the time. They simply comprised electrical heating elements that heated a plate, which had contact with the twisted yarn. Such heaters did not provide sufficient temperature uniformity along their length for a consistent heat-set process between machine processing positions, which is necessary for dye-uptake uniformity in the yarns.

In the mid 1970s, the introduction of vapour phase heaters (Dowtherm) to texturing machines brought about a significant improvement in yarn-temperature consistency and, as a result, textured yarn quality.[2] The operating principle of the Dowtherm heater is by heating a heating medium via an electrical heating element, the medium condensing onto the back of the metallic grooved plate which contacts the yarn. Thermocouples control the set temperature of the plate, which maintains good temperature consistency through condensate contact along its total length. This is demonstrated in Fig. 3.1.

Based on the experimental work concerning optimum residence time of a 167dtex commodity polyester yarn on a heater track (see Section 3.2), for desired yarn crimp contraction and stability to be realised at draw-texturing speeds of 800 m/min, texturing machines would require vapour phase heater lengths in the order of 2.5 m. In the case of nylon draw texturing,

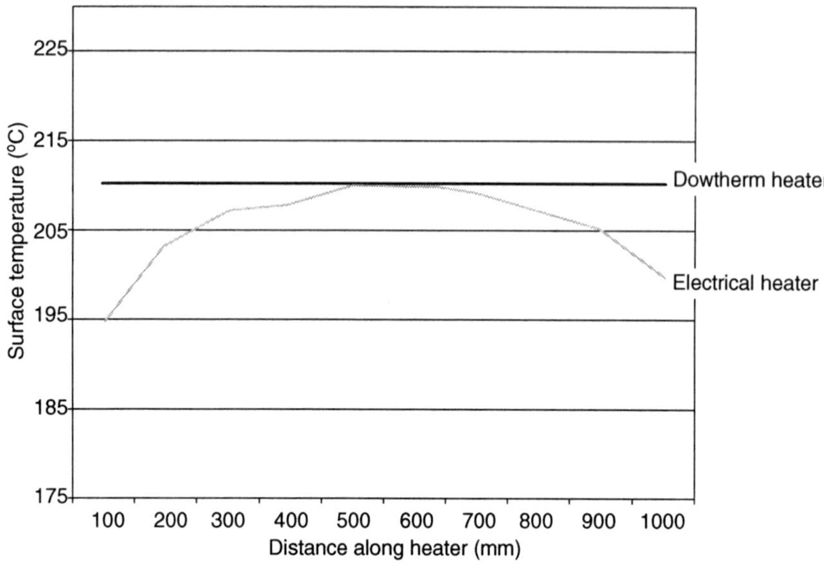

3.1 Temperature profiles for vapour phase and electrical heated plates.[2]

yarn count requirements for high elasticity end-uses, predominantly for hosiery, are significantly lower than with polyester. Therefore, 1.5 m to 2.0 m vapour phase heater lengths are quite capable of producing both satisfactory crimp-contraction and -stability. As a consequence, the draw-texturing industry experienced a new generation of machines in the early 1980s with vapour phase heaters, comprising track lengths of 2.5 m and 1.5 m–2.0 m for polyester and nylon applications respectively.

In practice, yarn temperatures when exiting heaters in draw texturing are generally in excess of 190°C for polyester and nylon applications. If it was possible to operate in the order of 1200 m/min process speeds for PET 167dtex, the above experimental results suggest that a heater length of 2.6 m would be required with a heater set temperature of 240°C. There are, however, fundamental factors, other than heat transfer to the yarn, that restrict maximum process speeds attainable in draw texturing; in reality, such speeds cannot be realised with conventional thread paths in the texturing zone on polyester 167dtex yarn.

Vapour phase heaters tend to be designed in demi-bay form, i.e. a housing block comprising six processing thread paths, the heater tracks of which are normally in dual U form (Fig. 3.2).[10] The tracks are normally ceramic plasma coated or nitride steel with an overall radius typically around 18–20 m. Conventional vapour phase heaters are used widely in the industry today. Although this heating technology has been used on draw texturing machines for around 30 years, they still have a secure future because of their ability

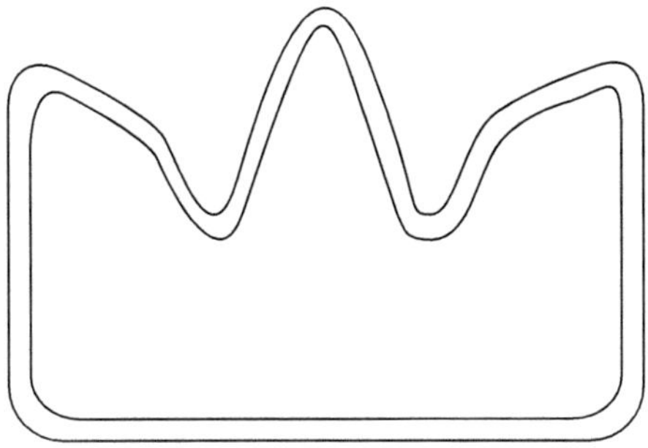

3.2 Cross-section example of a conventional heater track.

Table 3.2 Typical vapour phase heater lengths in use today

	Polyester	Nylon 66	Nylon 6	Polyproylene
Track length (m)	2.0–2.5	1.8–2.0	1.8–2.0	1.8–2.0
Typical temperature range (deg C)	190–230	190–220	170–190	110–180

to maintain tight temperature tolerances and consistency between processing positions. Typical vapour phase heater lengths used today and their set temperature ranges applied to various yarn types are given in Table 3.2.

3.4 High-temperature (HT) heaters

With the aspirations to produce polyester draw-textured yarns at ever increasing speeds and the associated increase in heater lengths that were deemed necessary at the time, it is not surprising that new heater technology emerged in the mid-1980s. It arrived around 1986 in the form of a high-temperature electrical heater, developed by Teijin Seiki, Japan.[11] The heater was 1.0 m long, with a twin temperature control zone. The technology at the time was believed to revolutionise the draw-texturing industry and other machine manufacturers rapidly followed with their high-temperature heater versions.

For PET application, HT heaters tend to comprise ceramic guides, which dictate the yarn path, supporting the twisted yarn; they are intermittently positioned along the heater length. The yarn is heated by convection through electrical heating elements located in close proximity to the yarn path. Typically, the heaters are sub-divided into two zones, with independent temperature control for each zone. Temperature range capabilities are usually up to 600°C (Fig. 3.3). Applied set temperatures are high, with HT heaters typically 380°C to 420°C for processing PET 167 dtex at 1000 m/min. Clearly, with such heaters, set temperatures are well above the melting point of the

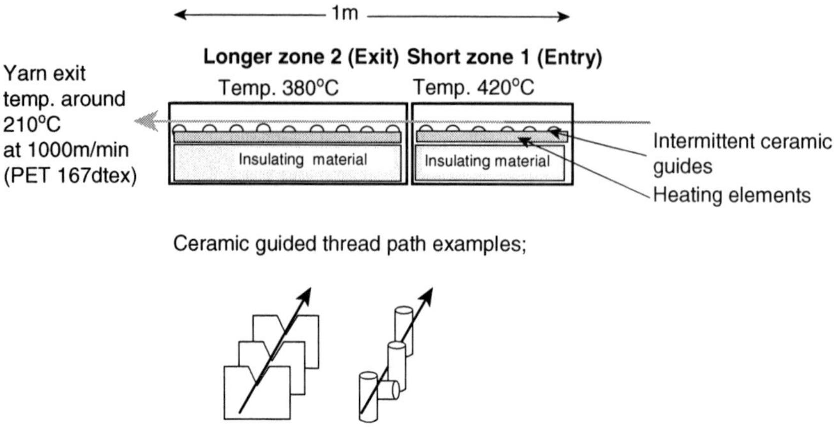

3.3 High-temperature heater application.

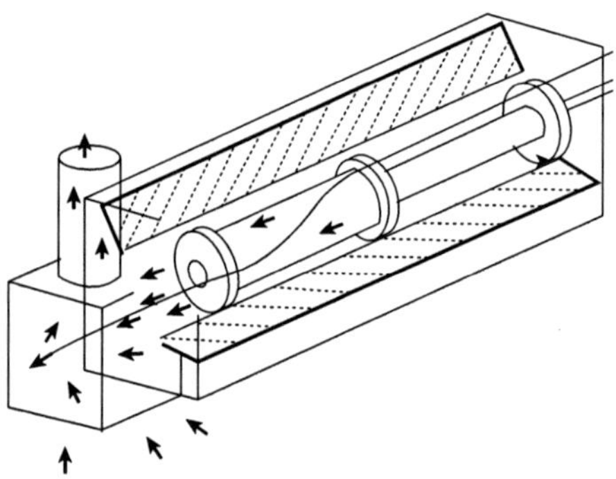

3.4 ICBT helical high-temperature heater.[12]

polymer, but for short yarn residence times, yarn temperatures exiting the heaters remain within a normal operational range.

There have been exceptions to the above guided HT heater design concept. For example, in nylon applications, Giudici, Italy, and Rieter-Scragg, UK, introduced contact HT heater tracks. In the case of Rieter-Scragg, these were plasma-coated metallic tracks ranging from 0.5 m up to 1.0 m for fine and medium denier nylon textured yarn applications respectively. ICBT, France, also marketed a cylindrical heated ceramic tube variant, predominantly for nylon texturing.[12] Here, the twisted yarn was heat-set through helical contact with the ceramic cylinder (Fig. 3.4).

HT heaters were predominantly introduced to the draw-texturing industry because process development work showed that by replacing the conventional 2.5 m vapour phase heater with a 1.0 m HT heater, the texturing zone length could be significantly shortened, enabling higher process speeds to be attained (Fig. 3.5).[11,35] This is because the maximum attainable speed,

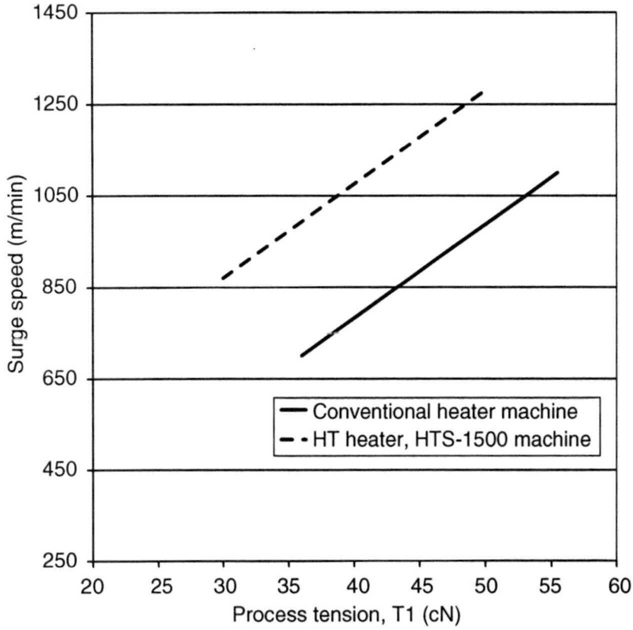

3.5 Surge speed advantage claimed by Teijin Seiki on their HTS-1500 texturing machine for the same twist level of 3400 turns per metre on a 167dtex polyester yarn.[11]

beyond which process instability, otherwise known as 'tension surging' occurs, is progressively increased with shortening of the texturing zone. Surging is discussed in more detail in Section 5.2.

HT heaters indeed have both advantages and disadvantages over the conventional vapour-phase heaters in draw texturing. These are considered under the following six headings:

(i) *Machine profile*: Because of excessive heater lengths, machines fitted with vapour phase heaters require large floor space and/or ceiling height. To overcome these limitations, some machine designs incorporate folded heat–cool zones (see Section 3.9.1), but these are limited regarding process flexibility needs on a wide range of yarn linear densities and filament counts. Machines with HT heaters require less floor space and can more readily be accommodated in a full range of machine profiles. For straight heat–cool zone thread paths (e.g. V-profile machines), a rotating twist stop is ideally necessary at the heater input to eliminate twist running back along the unsupported thread path to the input yarn feed (see Section 3.9.2). Twist stoppers are not operator-friendly, and when damaged can adversely affect yarn tensile properties and quality. The use of HT

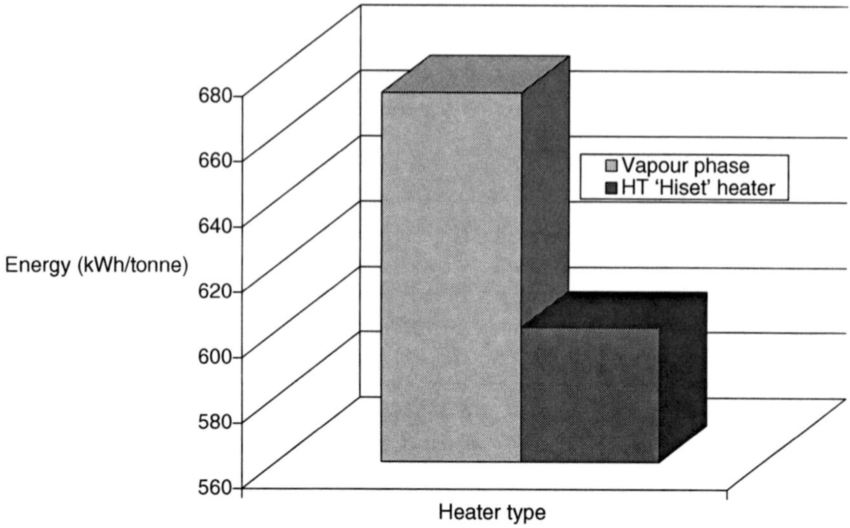

3.6 Reduced energy consumption using HT 'Hiset' heaters on Rieter-Scragg draw texturing machine for polyester yarns.[13]

heaters compliments the design potential for the input yarn feed to be in the vicinity of the heater entry, eliminating any need for twist stoppers.

(ii) *Ergonomics*: Due to reduced machine width and height, the threading of texturing zones with HT heaters is more operator-friendly than with the alternative longer vapour-phase heaters. This potentially improves production efficiencies and reduces mis-threads through improved operator efficiencies and reduced operator fatigue levels.

(iii) *Energy*: Machine manufacturers claim significant energy savings are realised for machines equipped with HT heaters (Fig. 3.6).[13]

(iv) *Maintenance*: In a production plant, conventional vapour-phase heaters are typically cleaned at a 4–6 weeks cycle period. Debris build-up on the heater tracks increases the yarn break rate, especially for the fine denier nylon yarns. In addition, debris can cause variation in heat transfer, which affects yarn dye uptake, increases the risk of operator mis-threading where the yarn is not fully located in the heater track and can be a cause of the yarn instantaneously rolling out of a track as a result of yarn ballooning due to yarn tension transient faults.[14] Typically, up to 2% output efficiency per machine per annum can be lost due to machine cleaning downtime, which is a significant cost factor when considering the process economics. Conventional vapour-phase heaters are cleaned by applying an appropriate oven-type cleaner at reduced heater temperature[15] and/or simply scraping the tracks with a brass wire brush. Abrasive wheel systems, driven by compressed air and fitted with a rotating brass brush are also available and prove to be very effective for heater cleaning.[16] Innovative, automatic cleaning methods by applying pulse laser beams to the heater during yarn running have also been explored in an attempt to reduce machine maintenance downtime but such methods seem to have not been further explored within the period of demise of the draw-texturing machine build industry in Europe.[17]

HT heaters were initially considered to be self-cleaning; simply by raising their set temperature to 600°C for a period of time, debris containing polymer, polymer additives and spin finish was initially found to burn off the yarn contact surfaces within the heater. Over lengthy machine runtimes, however, it was found that process break rates and output inefficiencies rose, due to hard deposit build-up on the heater guides. Normal cleaning methods using metal wire are not totally effective in removing these deposits. As a result, Barmag introduced interchangeable metallic sleeves to their twin zone HT heater. The sleeves are fitted with intermittent ceramic yarn guides, which can be cleaned remotely by ultrasonic means. Despite these deposit problems, heater cleaning cycles remain at least in the order of 4-fold higher than with conventional vapour-phase heaters for polyester yarn processing.

(v) *Yarn quality*: There are differences in the characteristics of yarns processed on vapour phase- and HT heaters. This is possibly due to friction, heat transfer- and filament deformation-differences within the twisted yarn core. It is difficult to match the exact conventional heater yarn aesthetics and optical performance in fabric with HT heaters. Polyester draw-textured products are, however, proven to be commercially acceptable for commodity yarns processed at the higher attainable speeds with HT heaters.

When using HT heaters, there is also evidence of a higher tendency to filament flattening on filaments located on the outside of the twisted bundle as a result of their contact with higher surface temperatures on the HT heater (Fig. 3.7). This can affect filament optics. The phenomenon can also be linked to a problem when processing nylon yarns. Here, at the high set temperatures necessary to generate adequate crimp values, together with the high yarn tensions required to support high-speed processing, inter-filament adhesion can occur between flattened filaments on the periphery of the twisted bundle, creating 'tight-spots' in the textured yarn.[18] Tight-spots can show as defects in knitted or woven fabrics (Fig. 3.8).

The operating window for optimising process specification (in terms of broken filament levels, process break-rate, and dye uniformity) tends to be lower with HT heaters than with the conventional

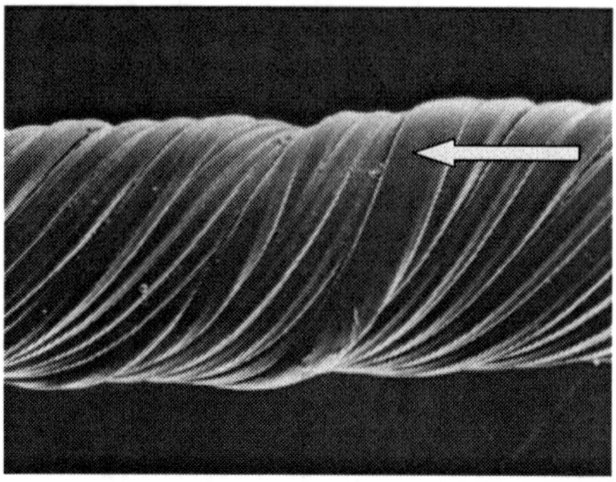

3.7 Flattened, plasticized filament on exiting heater in texturing zone.

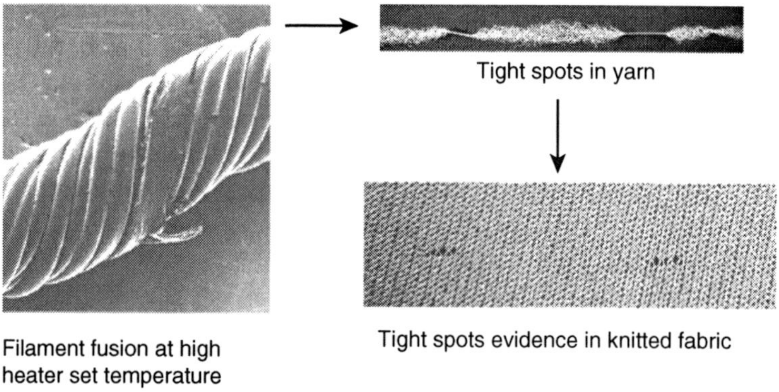

3.8 Nylon inter-filament fusion and resultant 'tight-spots' causing defects in knitted fabric at high set temperature with HT heater.[18]

vapour-phase heaters. Filament damage at high set temperatures, and sometimes the build-up of deposits on the yarn contact surfaces within the heater, can lead to deterioration in yarn tensile performance and to the onset of broken filaments. Also, for process conditions that lead to yarn ballooning between the yarn support guides within the heater or on exit to the heater, dye-uptake irregularities can lead to streaky fabrics.

The HT heater with intermittent ceramic support guides is not recommended for nylon yarns. Over relative short production runs, sticky nylon deposits tend to build on the ceramic guides, in close proximity to the running yarn. Any slight deviation in yarn path due to short-term tension fluctuations can cause the yarn to contact such deposits, leading to breakage of the yarn. In practice, process break rates have been seen to increase four-fold with some nylon processes due to this phenomenon over a few weeks of continuous running.

Basically, the HT heater with ceramic yarn support guides should be viewed with considerable caution for nylon draw-texturing applications.

(vi) *Process flexibility*: Draw-texturing companies, which require access to numerous feedstock yarn suppliers for yarn price or yarn type reasons, tend to prefer to texture using vapour-phase heaters. These provide a wider window for optimising process specifications, and are not as sensitive as HT heaters to spin finish type. Similarly, where high process speeds are not realistic for fancy yarn production or micro-filament, vapour-phase heaters show lower tendency towards the onset of filament damage and broken filaments, and are the preferred choice.

3.5 Intermediate-temperature heaters

Within the last five years, some machinery and component manufacturers have reverted attention to electrical resistance heated tracks, along which the yarn has full contact. The basis of these developments is to address the shortcomings of the HT heater, whilst compromising on heater length. Examples here are Dienes (Germany) and SSM (Switzerland) on their new SSM DP3-FT false twist texturing machine.[18,19] In both cases, the heaters are 1.5 m in length and comprise triple (Dienes) or twin (SSM) zone electrical resistance heaters. Good temperature consistency along the heater length is claimed through element design. In addition, the heater radii are designed in such a way that instability process speed is not compromised compared to the HT heater.

Wide production experience with such heater designs is, however, limited to date and is believed to be purely with nylon, where the HT heater has extremely limited application. To attain adequate crimp levels in draw-textured yarns, typical set temperatures with these heaters are around:

- Polyester, 167dtex, 1000 m/min, 280°C.
- Nylon 66, 78dtex, 900 m/min, 230°C.

The same heater cleaning methods are applicable to 'Intermediate Temperature' heaters as those used with vapour-phase heaters.

3.6 Comparison summary of heater types

Taking all the factors discussed in Sections 3.2 to 3.5 into consideration, ideal heater selection for yarn types, yarn counts and filament fineness can be summarised as in Table 3.3. The table clearly shows why the 'straight' heat–cool thread path with conventional 2.0 m heaters remains the norm

Table 3.3 Heater selection for yarn types, yarn counts and filament fineness

Issue	Vapour phase: Folded yarn path	Vapour phase: Straight yarn path	Straight yarn path: HT non contact (ceramic guides)	Straight yarn path: HT full contact (track)	Straight yarn path: Intermediate temp. (track)	Key and comments
Heater length (m)	2.0–2.5	2.0	1.0	0.5–1.0	1.5	Good / Satisfactory / Poor
Floor area	Good	Good	Satisfactory	Satisfactory	Satisfactory	
Ceiling height	Poor (2.5 m length)	Good	Satisfactory	Satisfactory	Satisfactory	
Ergonomics	Good	Good	Satisfactory	Satisfactory	Satisfactory	
Energy	Good	Good	Satisfactory	Satisfactory	Good ?	
Maintenance (cleaning)	Good	Good	Satisfactory	Satisfactory	Good	
Polyester processing:						
Speed/efficiency:						Tendency to filament damage on folded thread path with yarns
Commodity yarn	Satisfactory	Good	Satisfactory	Poor	Poor	
Microfilament	Poor	Good	Good	Poor	Good	
Product quality:						Insufficient heat transfer for short HT contact heater within realistic temperature operating range
Commodity yarn	Good	Satisfactory	Good	Poor	Good	
Microfilament	Poor	Satisfactory	Good	Poor	Good	
Process flexibility	Poor	Satisfactory	Poor	Poor	Good	
Nylon processing:						
Speed/efficiency:						High break rates on HT heater
Medium yarn count	Good	Good	Poor	Poor	Good	
Fine yarn count	Poor	Satisfactory	Poor	Good	Good	
Microfilament	Poor	Satisfactory	Poor	Good	Good	
Product quality:						Tendency to filament fusion with HT guided heater and filament damage on folded thread path
Medium yarn count	Good	Satisfactory	Poor	Good	Good	
Fine yarn count	Poor	Satisfactory	Poor	Good	Good	
Microfilament	Poor	Satisfactory	Poor	Good	Good	
Process flexibility	Poor	Satisfactory	Poor	Poor	Good	
Polypropylene processing:						Temp. range, 110–180 deg C required for process flexibility
Speed/efficiency	Poor	Good	Good ?	Poor	Good	
Product quality	Poor	Good	Good ?	Poor	Good	
Process flexibility	Poor	Poor	Good ?	Poor	Satisfactory	

in today. The introduction of folded heat–cool zones with vapour-phase heaters and subsequently HT heaters was aimed, at the time, at higher process speeds and reduced energy costs, to reduce the yarn conversion costs in production. They have not, however, fully satisfied the increasing need for processing wider product ranges (yarn counts, filament fineness), which can be accommodated by the vapour phase heater in a 'straight' heat–cool zone profile. Recent developments with intermediate temperature, electrical resistance heaters may be of interest in the future in terms of process flexibility and energy saving. Oerlikon-Barmag, for example, has also reverted to electrical resistance heaters, utilising full yarn contact heater inserts, for energy saving on finer count yarns.[20] The effect of texturing zone thread path wrap angles on yarn quality and process speed limitations is discussed further in Sections 3.9.1 to 3.9.3.

3.7 Yarn cooling in the texturing zone

Technology advance in heater design has contributed to the increase in process speeds attained in draw texturing. As process speeds increase, however, more demands are imposed on cooling the hot, twisted yarn before it is de-twisted on exiting the false twisting device. At the very start of draw texturing, cooling was carried out simply by running the unsupported yarn through an air gap at room temperature. This could be considered satisfactory at the early, extremely low process speeds. However, as speeds increased, it was necessary to run the yarn over a support track to avoid yarn ballooning. During texturing, twist is imparted to the yarn at extremely high rates; for example, in the order of 32 000 and 105 000 turns per second for polyester 167dtex and nylon 17dtex yarns respectively. Without support and stabilisation, the twisting causes the yarn to balloon, which brings about tension fluctuations and resultant dye shade variation in the fabric. Moreover, it reduces the maximum process speed attainable.

Also with increase in process speeds, more stringent cooling requirements are found to be necessary. Here, requirements seem to be different for polyester and nylon processes, possibly due to their differences in behaviour around their glass transition temperatures (Tg):

- Polyester, Tg = 70–90°C.[37,38]
- Nylon 66, Tg = around 50°C.[34]
- Nylon 6, Tg = around 47°C.[34]

Other literature, however, reports that PET polyester and nylon 6 have similar glass transition temperatures around 100°C.[39] This literature also explains the more complex 'glass to rubber' transitional state associated with these two polymers.

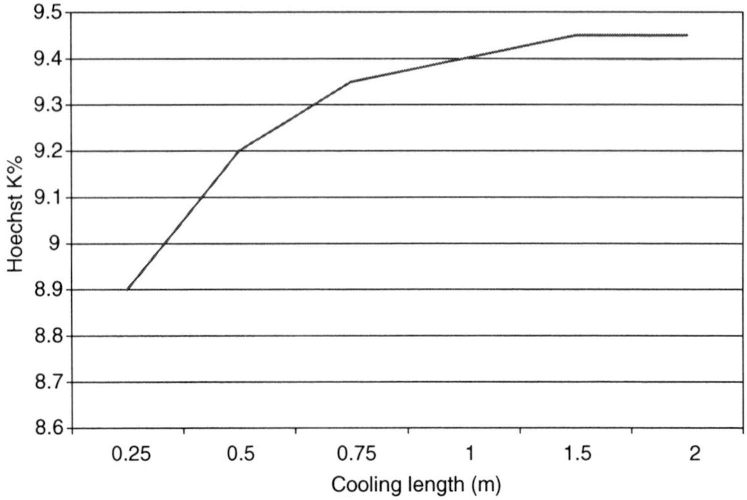

3.9 Effect of cool track length on crimp value for polyester 167decitex yarn.[21]

The glass transition temperature of a polymer is the temperature at which it changes from a rubbery state to a glassy form. When in a glassy form, in response to tension, elastic materials will return to their original shape on removal of the force. For temperatures well above the Tg, repeat units of the polymer structure are relatively free to move, whereas below the Tg, the motion of these individual polymer chain segments become frozen with only small-scale molecular motion remaining. PET polyester seems to be more susceptible to molecular restructuring under applied load above its Tg it would appear to be necessary to cool PET polyester yarns sufficiently, so that their temperatures on de-twisting do not exceed the Tg, otherwise slight tension variation between processing positions can lead to variation in molecular structure leading to dye-shade variation. This in turn may cause defective fabric.

The cooling process in draw texturing was extensively investigated during the 1970s and 1980s:

- For 167dtex polyester yarns, crimp contraction was found to increase before reaching a plateau with increase in cool plate length (Fig. 3.9).[21] Similar trends were evident with a nylon 78dtex yarn at a fixed process speed (Fig. 3.10), with steady crimp values being realised over a relatively shorter cool length than with the heavier polyester 167dtex yarn.
- The rate of change in dye uptake reduced with increase in cool-track length, and tended also to plateau with no further change when reaching plateau conditions for crimp at a fixed process speed.

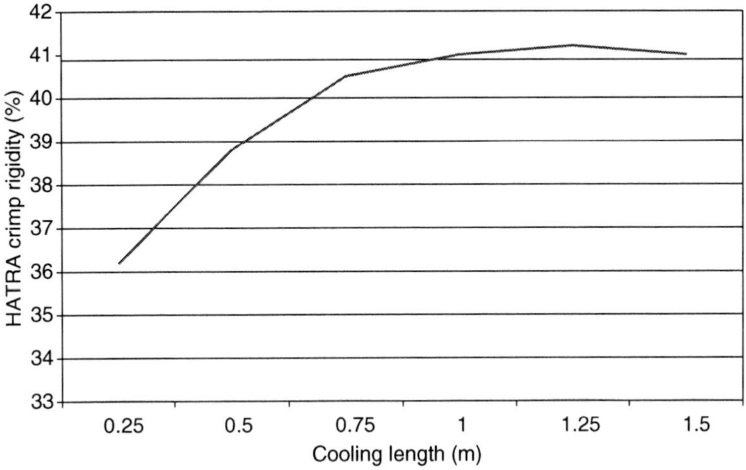

3.10 Effect of cool track length on crimp rigidity values for nylon 66 78 decitex yarn.[21]

- For short cool lengths below that required for stable conditions, yarns dyed darker with reduction in cool length.
- Water-cooled tracks were introduced for a short period to the Scragg SDS7 in the early '70s for texturing polyester. The concept was not successful, possibly due to inadequate temperature control of the coolant.

Cool track lengths exceeded 2.0 m in the early 1980s for polyester draw texturing applications in order to satisfy the above criteria. However, when these cool track lengths are combined with 2.5 m vapour-phase heaters, ergonomic, floor space and machine height problems materialised as texturing zone lengths reached approximately 4.8 m. For such designs, texturing zone lengths exceeding 4.8 m were common. As a result, new machine profiles emerged, largely for polyester yarn applications, with thread path wrap angles between heater and cool track. (These are discussed in Section 3.9.1.)

The effect of texturing zone profile on yarn entry temperature to the twist unit has been reported for commercially available machines.[22] Differences in entry temperature due to the different cool track lengths between straight and folded heat–cool zones are marked, potentially limiting the maximum speed attainable on the straight zone variant for high decitex polyester yarns (Fig. 3.11).

In practice, cool plate lengths for the 3rd to 4th machine generations (see Table 3.1) were commonly between 2 m and 2.5 m in length to bring about adequate cooling of polyester yarns at modest process speeds.[22,23] For polyamide applications, however, relatively shorter cool track lengths were found to be acceptable, typically 0.7 m to 1.0 m. As a result, texturing

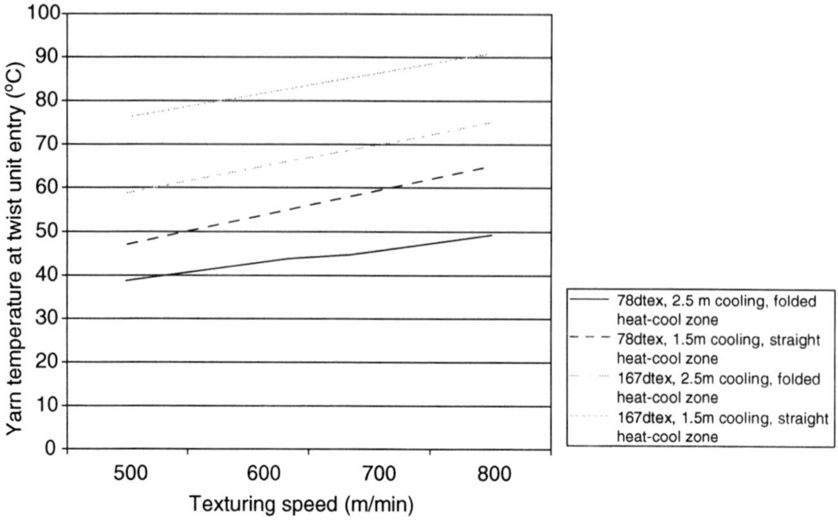

3.11 **Differences in yarn temperature at entry to the twist unit due to different cool track lengths on straight and folded heat–cool zones.**[23]

heat–cool zones tended to remain straight within the machine evolution for this particular end-use.

There seems to be some conflict today concerning past findings on cool track lengths; modern draw texturing machines are claimed to be able to texture 167 dtex yarns at process speeds in excess of 1000 m/min with cool track lengths less than 1.7 m. Improved consistency in process tensions due to improved twist unit and friction disc designs and possibly better POY quality for more uniform process tensions in texturing may be contributory factors. Nevertheless, there are some indications that inadequacies in cool track designs and length for more recent machine designs may still be consideration points by machine manufacturers. Cooling tubes enabling a helical yarn path were developed in 1997 (Fig. 3.12).[24] This brought about more effective cooling due to increased yarn residence time on the cooling surface due to the helical yarn path, enabling shorter cooling zone lengths in machine design. The concept is evident today on some recent machines produced for polyester applications.[20]

There has been further development of cooling tubes, incorporating an adjustable helix angle which allows change to the effective angle of yarn wrap on the tube, thus positively affecting process instability speed and hence the maximum process speed attainable.[25,26] There has also been the development of active cooling through direct yarn contact with a cooling fluid (water), which ensures that polyester yarn temperatures remain well below the upper limit at twist unit entry for high texturing speeds (Fig. 3.13). As a result, the total heat–cool zone length on a draw texturing machine

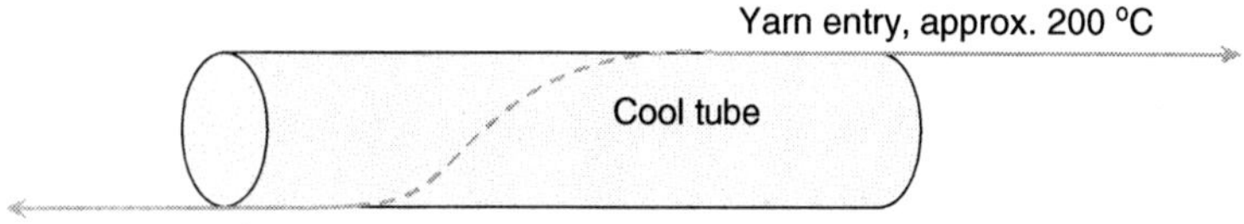

3.12 Cooling tube with helical yarn path for polyester applications.

3.13 'Temcooler' active water cooling system fitted to a production draw texturing machine.[27]

can be reduced by an order of 50%, allowing more compact machine designs and the opportunity to maintain process stability at higher texturing speeds.[27]

Of the above developments, active cooling can be considered the most innovative means of cooling the twisted yarn. The concept involves the yarn passing through an enclosed water bath, through which there is a metered flow of water at a steady temperature. The water is not carried forward by the yarn due to a purposely-designed compressed air labyrinth system at entry and exit to each cool box. As residual spin finish is removed from the yarn by the coolant, the system incorporates a means of controlling the spin finish concentration in the circulating coolant over the entire draw texturing machine. Although this cooling method stimulated much interest over the last several years, it seems to have been commercially introduced only on a very limited scale. This is probably mainly due to:

- The complexity of the system and investment costs.
- Higher process speeds in excess of 30% needing improved heater technology.
- The rapid demise of the draw texturing industry in the Western World and its relocation to the Far East and Asia, where the low capital investment cost of locally produced standard machines already offsets

the process cost advantages of the higher speeds claimed with the active cooling system.

3.8 Secondary heaters

For textile applications where reduced elasticity (and reduced residual yarn twist liveliness) is required, e.g. polyester yarn for weaving, it is necessary to relax the textured yarn in secondary heaters that are integrated into the texturing machine as a secondary thermal treatment after the texturing zone. Secondary heaters are 'non-contact', i.e. in tubular form with limited surface contact on the running textured yarn. Their designs ensure limited yarn contact friction, as extremely low yarn tensions are necessary within this zone to enable the stresses imposed on the yarn in the primary texturing zone to relax.

There are two process parameters in the secondary heater zone, which can be changed to affect crimp and residual torque in the yarn:

- Heater temperature: increasing the heater temperature reduces the crimp value and residual torque of the yarn. There are, however, maximum temperature thresholds beyond which streaks and flecks are evident in the yarn after dyeing, due to temperature and tension variation effects in the zone.
- Yarn overfeed: it is normal to operate with high yarn overfeed into the heater zone to allow the molecular stresses to relax. Again, however, there are thresholds. The maximum overfeed is limited as high yarn flutter, especially at high set temperatures, can cause dye flecks due to intermittent yarn contact with the heater tube surface. Also, when tensions are too low at the heater entry as a result of high overfeed use, any short-term tension disturbance from the primary texturing zone can cause sufficient disturbance in the secondary heater zone for the low tension yarn to wrap and break on the delivery system to the heater entry.

Heater designs can be electrical or vapour phase, with set temperature ranges typically between 130°C to 240°C, each yarn end passing down a stainless tube (ca. 3–5 mm diameter), which is removable for cleaning purposes. Heater lengths range from 700 mm to 1500 mm. Internal tube variants are available in terms of helical protrusions, which are claimed to reduce filament loop tendencies on set multi- and microfilament yarns.[28]

Over the last three decades, there have been some exceptions and variants to the conventional, vertical thread path secondary heater designs:

- Murata's high temperature heater, with a temperature range from 100°C–600°C and open threading.
- Rieter-Scragg's warm air injection system for improving heater efficiency (Table 3.4).[23] Warm air is injected into the heater tube to reduce

Table 3.4 Lower polyester yarn bulk achievable with air assisted, high efficiency heater[23]

Secondary heater type	Polyester yarn type	Maximum set heater temperature above which dyeshade irregularities are apparent	Minimum crimp level (Hoechst K%), which is achievable at this maximum temperature
Conventional tube	167dtex f34	215	24
	78dtex f34	215	18
Air assisted tube heater	167dtex f34	215	14
	78dtex f34	215	7

the inefficiency caused by cold air being dragged into the tube by the yarn. The system is also claimed to enable higher yarn overfeeds due to the forwarding action of the air. This enables lower yarn crimp values to be realised, thus widening the process operating window for the secondary heater.

- RPR's heated godet systems.
- Barmag's angled heaters on their MPS machine.[30]

It is difficult to believe that some of the above variants do not limit the full overfeed and temperature operating windows that can be applied to the draw textured yarn for the reasons outlined above in the brief discussion on the effect of parameters. For ladies' hosiery, with increase in yarn variants for fashion trends and the increased use of covered elastanes for elastic form and comfort, limited use of secondary heaters has also evolved in recent years in the texturing of nylon yarns.

3.9 Machine profiles

Research in the 1970s to 1980s showed that a heater residence time in the order of 0.2 sec. must be applied to polyester texturing for optimum conditions to be realised (Section 3.2). Moreover, in subsequent cooling, the detwisting point on the exit to the twist unit should be at a temperature below the glass transition of PET for good dye shade uniformity. As a result, the industry saw the emergence of texturing heat–cool zones in excess of 4.5 m in length for process speeds up to 800 m/min.

To accommodate limitations in room space (height and width) and to ensure good ergonomics, folded heat–cool zones were introduced by leading machine manufacturers over an interim period, e.g. Barmag M and Rieter-Scragg B profiles in the 1980s (see Folded Thread Path, Fig. 3.14). The concept necessitated ceramic guides between the heater exit and cool track

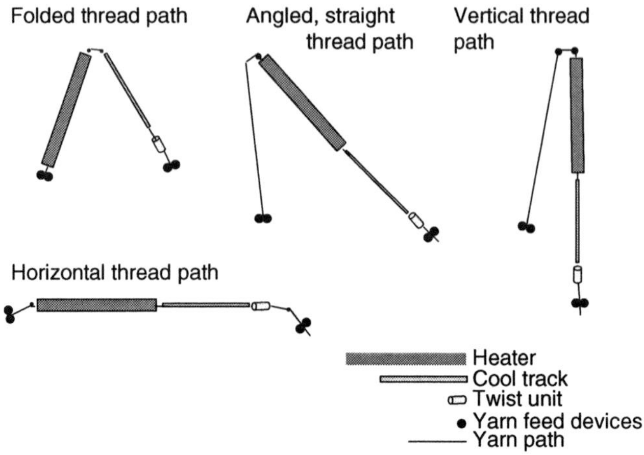

3.14 Examples of common texturing zone profiles.

entry to support the twisted yarn. This ceramic guide system was built into a support sledge, which could be drawn down for threading and raised back to its docking location for yarn processing.

Subsequent development of finer filament yarns (low d.p.f. – denier per filament), which are sensitive to filament breakages, and the need for more process flexibility to accommodate an ever-increasing range of yarn types, brought about the rapid reintroduction of straight heat–cool zones (see Vertical, Horizontal and Angled Thread Paths, Fig. 3.14). Such texturing zone thread paths are generally preferred today. The advantages and disadvantages of the various designs of texturing zone profile are discussed in Sections 3.9.1 and 3.9.2.

3.9.1 Folded thread path

Folded heat–cool zones were introduced on a large-scale to polyester draw texturing in the early 1980s. The concept was designed largely for the texturing of the following yarn types, 78 dtex f34, 167 dtex f32, 34 or 36 and 330 dtex f64, 68 or 72. These yarns constituted the bulk volume of the polyester draw textured yarn market at the time and were mainly for weaving applications.

Even though the folded heat–cool zone tends not to be prevalent in machine sales today, it must not be forgotten that the concept was designed at a time when there was much international R&D activity in draw texturing. Moreover, there remains evidence of such machine types still operating worldwide, largely on low sensitive, commodity yarns. The experiences gained in both development and plant operations within that machine era have, to a large extent, provided a platform for some of the more recent developments evident in draw texturing machine design. Indeed, there are some advantages of the folded thread path concept that should not be ignored in future machine designs (see Process instability tension transients and speed limitations due to 'surging', Section 5.2). The success of the folded texturing zone was very much dependent on the surface, thread line geometry and diameter of the two ceramic guides located between the heater exit and cool track entry.

Surface roughness of the guides and total angle of yarn wrap can be considered the most critical parameters for optimum texturing performance, as they influence:

- Torque transmitted to the twisted yarn on the heater due to surface friction.
- Filament migration characteristics in the twisted bundle.
- Yarn twist level on the heater and crimp values: twist transmission to the heater is adversely affected by friction at the turn-round guides (Fig. 3.15) and yarn tensions *T1* and *T2*, i.e. tensions before and after the twist unit increase significantly due to friction at the turn-round guides. For example, T1 tension prior to the twist unit can be expected to be in the order of 20–25% higher than on straight heat–cool zones for a PET 167 dtex yarn at the same applied draw ratio.
- Yarn tensile performance and number of broken filaments (Fig. 3.16).

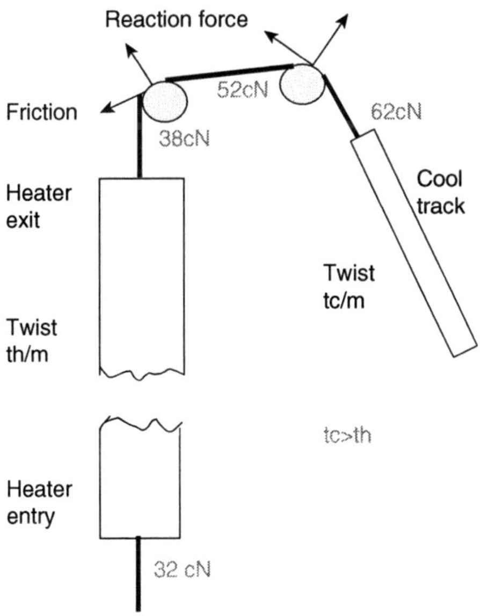

3.15 Example of tension build-up due to friction at turn-round ceramic guides on folded heat–cool zone (Polyester 167dtex f34 yarn).

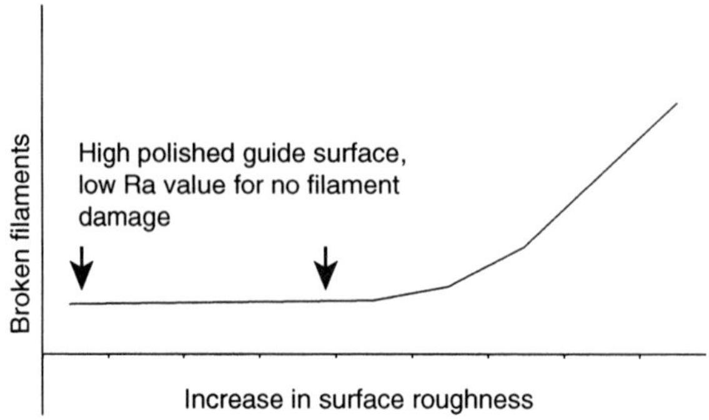

3.16 Increased tendency to broken filaments with increased surface roughness of ceramic guides between heater and cool track.

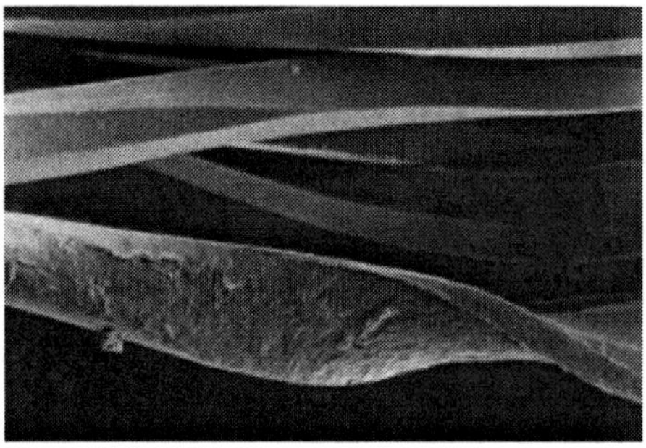

3.17 Evidence of filament flattening on filaments that were on the outside of the twisted bundle.

For the high process tensions that are required to support high-speed operations, at the ceramic turn-round guides reaction forces on individual filaments on the periphery of the twisted bundle are high. Moreover, the filaments are still at high temperature on exiting the heater and the forces tend to enhance a tendency to filament flattening (Fig. 3.17).

During twisting, the filaments tend to migrate within the twisted bundle and exposure to flattening concerns filaments that are, for a short time, on the outside of the bundle. These are weakened and vulnerable to breakage, particularly during the stresses imposed on the yarn during de-twisting on the exit to the twist unit (Fig. 3.18).[31] This is a particular problem when processing yarns with finer filaments, especially micro-filament yarns. With folded texturing zone thread paths, it is common to operate with a relatively higher number of friction discs on the twist unit to reduce *T1* tension (increase the twist level) to offset the loss in twist on the heater due to friction at the turn-round and to improve filament migration at the heater entry through increased torque.

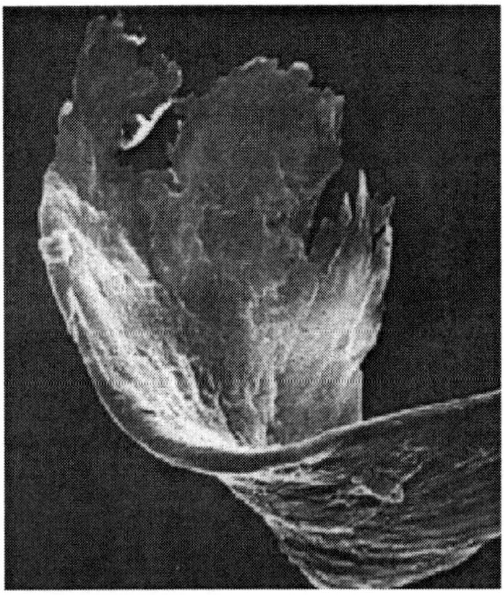

3.18 Broken, flattened filament.[31]

Importantly, the ability to operate with the input yarn feed in close proximity to the heater entry, avoids excess unsupported yarn lengths, which would be susceptible to 'ballooning', resulting in tension variations. Such tension fluctuations can reduce the maximum process speed attainable through the onset of process instability or 'surging'. Of particular importance too is the use of additional ceramic guides in this pre-heater thread path; for problematic yarns this can facilitate improved twist migration, increasing crimp frequencies in the filaments and reducing tendency to filament loops and damage (Fig. 3.19). This is particularly advantageous for yarns where inter-filament friction levels are high due to spin finish properties, or in instances of difficult filament migration due to high filament numbers. A major advantage of the folded thread path is that for specific draw ratios applied to PET yarns, the maximum process speed attainable due to the onset of 'tension surging' is higher. This is because of the lower twisting torque levels experienced by the yarn on the heater and the higher yarn tensions associated with its thread path design. Process instability is discussed in Section 5.2.

3.9.2 Straight thread path

Heat–cool zone lengths that are optimum for high-speed polyester yarn texturing cannot be completely applied for angled, horizontal, or vertical texturing thread paths due to inferior ergonomics and the excessive floor space and ceiling heights required. However, the ever-increasing need to accommodate new yarn- and filament-linear densities has steered the

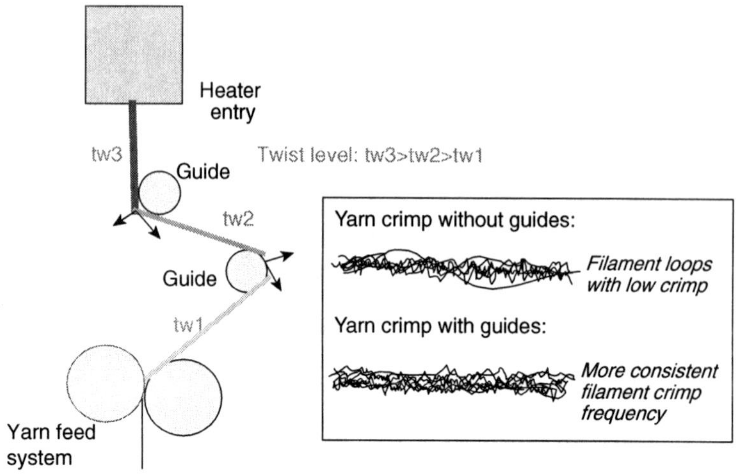

3.19 Facilitating filament migration in the twist through a double ceramic guide system at heater entry.

move back to straight heat–cool zones, with compromises being made on cooling length. Process speeds are limited for many of the more sensitive yarn types that have emerged and the high speeds that were being strived for on the basic commodity yarns are no longer a sole priority for the machine manufacturers. As a result, shorter cooling lengths could be accommodated and the straight heat–cool zone tends to be prevalent in the industry today.

To avoid twist run-back to unsupported yarn lengths between the input feed and heater on the angled and vertical heat–cool zones, rotating twist stoppers are used (Fig. 3.20). Their design consists of opposing plates with teeth, which provide sufficient friction to stop twist run-back. In the past, stationary guide systems have also been tried for this application, but their use is limited, as to stop the twist run-back, large angles of contact wrap are necessary, which lead to very low tension upstream of the guide system to the yarn input feed and thus severely restrict the process tensions that can be applied.

Twist stops eliminate yarn ballooning prior to the heater; resultant tension fluctuations would otherwise reduce the surge speed due to these fluctuations (Table 3.5).[32] Twist stops are necessary but reluctantly used in draw texturing plants:

- They are generally manufactured in ceramic, which can be easily damaged or broken if dropped.
- Yarn wraps tend to occur on the twist stop when a yarn breaks in the texturing zone; for operators, it can be tedious and time-consuming to remove such wraps.

3.20 Rotating twist stop.

Table 3.5 Higher process instability speeds with use of twist stop[32]

Yarn type	Draw ratio	Process instability speed (m/min)	
		With twist stop	Without twist stop
110dtex f34	1.36	920	800
78dtex f34	1.35	970	930

- They are usually located in close proximity to the heater entry and their bearings are exposed to elevated temperature and spin finish fumes. As a result, good plant discipline is required to ensure that all units are freely rotating.
- High twisting torque at twist run back to the twist stop teeth can sometimes result in broken filaments when processing highly sensitive microfilament yarns. For this reason, it is sometimes beneficial to interchange the twist stop with its pre-support guide at the heater entry to enable guide surface friction to reduce the twisting torque on the teeth of the twist stop. In this case, however, a highly polished ceramic guide surface is required to avoid too high a yarn friction on the guide surface, which itself would otherwise cause filament damage!
- There is a cost element for worn or damaged twist stop replacements, which adds to maintenance expenditure.

As in any false twist texturing process, the maximum process speed is limited by the onset of process instability. Higher draw ratios must be applied to a straight heat–cool texturing zone to attain the surge-speed level of a folded thread path, the extent of which is limited largely by the minimum acceptable elongation of the textured yarn and indeed its linear density.

Also, there is a maximum draw ratio that can be applied due to the onset of broken filaments at a yarn tension threshold.

For similar crimp values, straight heat–cool zone machines tend to produce a slightly mossier appearance in knitted fabric than those with a folded thread path. This is most probably related to differences in filament migration characteristics at the initiation of twist before the heater, i.e. due to differences in twisting torque on entry of the yarn to the heater, and yarn temperature differences at exit to the twist unit.

On angled heat–cool zones, the threading of the heater track with yarn is carried out by use of a suction gun and a sledge assembly that supports the twist stop. The sledge is raised from the input feed after threading to its docking position at the heater entry. Raising the sledge used to be carried out manually with a lightweight pole, but now it is activated by pneumatic devices to improve operator efficiency and reduce operator fatigue.

Today, the adoption of the angled, straight heat–cool zone as the norm in false twist texturing machine design has brought about wider use of the cool tube concept to overcome the shortcomings of cool track length in the processing of polyester yarns.

Lowering the heat-cool zone into a horizontal or near horizontal profile:

- Widens the machine, utilising more floor space (cost factor!), but reduces machine height.
- Tends to be more demanding on the operator due to the wider service aisle between input feed and twist unit.
- Provides easier access to the heaters for cleaning at service intervals.
- Eliminates the need for rotating twist stops.

Clearly, short high-temperature heaters can be well integrated into the straight heat–cool zone design concept as:

- Better surge speeds and hence higher attainable process speeds can be achieved.
- Improved ergonomics and reduced floor space are realisable.
- There is a better opportunity to adopt a horizontal thread path, allowing the input yarn feed to be in close proximity to the heater, eliminating the use of twist stops.

However, as discussed in Section 3.4, HT heaters have limited process-operating windows and are restricted in satisfying the increasing process flexibility requirements of today's market.

3.9.3 Hybrid texturing zone profiles

There have been numerous attempts by machine manufacturers to consolidate the advantages of folded and straight heat–cool zones into

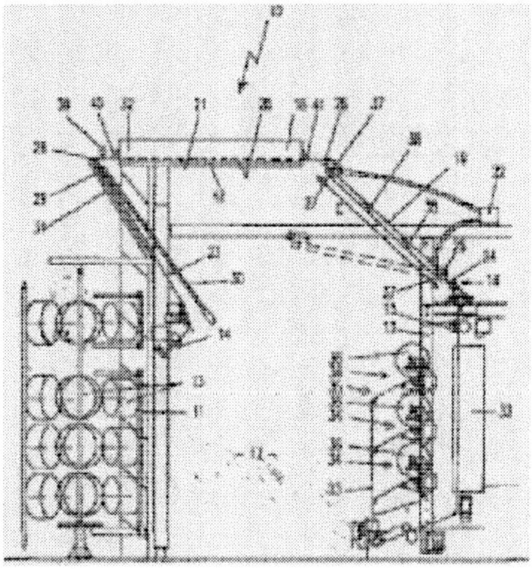

3.21 High speed texturing concept from Rieter-Scragg.[24]

single-thread path geometry, even including a novel rotating guide between heater and cool track, which allows twist transmission to the heater and reduces the yarn tension at the twist unit to that approaching a straight thread path.[33] A variant, however, of particular interest is a profile, patented by Rieter-Scragg, that incorporates the advantages of a horizontal heater, higher process surge speed capability and improved yarn cooling (Fig. 3.21).[24] This variant, aimed at the false twist texturing of polyester, potentially offers the following advantages:

- Slight thread path angle between heater and cool track to reduce aisle width and increase process instability speed.
- Horizontal heater for reduced energy consumption.
- Use of cool tube for better yarn cooling over a shorter distance.
- Adjustable yarn helix angle on cool tube for enhanced surge speed setting.

This texturing zone profile still, however, necessitates the use of twist stops. Due to the demise of Rieter-Scragg, there is no evidence of this concept becoming a commercial reality.

3.9.4 Summary of machine profile applications

As a guideline, the summary shown in Table 3.6 outlines the advantages and disadvantages of the machine texturing zone profiles available on the market today.

Table 3.6 Guideline for machine profile advantages and disadvantages

Parameter	Conventional heater				High temp. heater	
	Folded	Straight angled	Straight vertical	Straight horizontal	Straight angled	Straight horizontal
Max. process speed						
Process flexibility:						
Commodity PET						
Fine filament PET						
PA6 and 66						
Fine filamentPA6 and 66						
Polypropylene						
Process operating window						
Yarn physical properties						
Yarn tensile properties						
Threading (start-up)						
Ergonomics						
Floor space						
Ceiling height						
Ease of heater cleaning						
Incorporates twist stop?						

Key: satisfactory → poor

3.10 References

1 D. K. Wilson, The Production of Textured Yarns by the False-twist Technique, *Textile Progress*, Vol. 10, 33 (1978).
2 *Draw Textured Yarn Technology*, Monsanto Textile Company (1974).
3 H. M. El-Behery, Parameters of Existing Textured Yarn Processes and their Limitations, *False-twist Textured Yarns Workshop*, Charlotte, N. Carolina, March 23–24 (1972).
4 M. J. Denton, *Evolution in False-twist Texturing Machine Design, Developments in Texturing*, Shirley Institute Publication 525, Oct. (1976).
5 J. Parnaby, False-twist Advance, *Textile Asia*, 52–55, Feb. (1981).
6 C. Atkinson, J. Parnaby *et al.*, MMF Processing: The Balance Between High Speeds and Yarn Quality in Draw Texturing, *International Textile Machinery*, 35–38 (1981).
7 R. Machatschke, S. Mueller-Probandt *et al.*, Innovation in Texturing, *Melliand International*, 268–269, Vol. 6, Dec. (2000).
8 D. S. Barnes, W. J. Morris, Rates of Setting in False-twist Draw-texturing, Part 1: The Effects of Heating Time and Length in the Processing of Polyester Fibre Yarns, *Journal of the Textile Institute*, No. 6, 291–298 (1980).
9 D. S. Barnes, W. J. Morris, Rates of Setting in False-twist Draw-texturing Part II: The Effects of Heating Time and Heater Length in the Processing of Nylon 66 Yarns, *Journal of the Textile Institute*, No. 6, 299–304 (1980).
10 *Technical Literature*, Rieter-Scragg Ltd, 1981.
11 F. Tanae, Neue Strecktexturiermaschine HTS – 1500, *Chemiefasern Textilindustrie*, 40, 991–993, Oct. (1990).
12 J-C Dupeulle, Energy Saving using the Short Heater, *TYAA Annual Meeting*, Jul. 31 (1992).
13 *Technical Data Publication SDS900*, Rieter-Scragg Ltd. (1984).
14 H. M. Familant, Theory on Heater Cross-over in False-twist Texturing Operation, *Textile Research Journal*, 335–340, Jun. (1980).
15 *Technical Literature, Polyfix 600*, Schill und Seilacher, Germany, 1982.
16 *Pulitrice a Spazzola*, Technical Manual, COBRA, Calstel Goffredo, Italy, 2010.
17 D. C Eaton, C. Atkinson *et al.*, *Cleaning Textile Machines*, Rieter-Scragg Ltd. EP0853148 (AZ), 07.15.1998.
18 C. Atkinson, S. Müller-Probandt, Opportunities for Further Optimisation of the Primary Zone in False Twist Texturing, *Man-made Fiber Year Book 2006*, Chemical Fibers International.
19 *Technical Brochure DP3-FT*, SSM Schweiter Schärer Mettler AG, Switzerland, 2007.
20 U. Wagner, *MPS Coolflex – The Solution for Fine Denier Texturing*, Oerlikon–Barmag Corporate Communications Literature (2007).
21 D. S. Barnes, W. J. Morris, Rates of Setting in False-twist Draw-texturing Part III: The Effect of Cooling Time and Length in the Processing of Nylon and Polyester Fibre Yarns, *Journal of the Textile Institute*, No 6, 305–312 (1980).
22 K. H. Bauer, Herstellung neuer Garntypen auf Falschzwirn- und Lufttexturiermaschinen, *Chemiefasern Textilindustrie*, 40, 981–985, Oct. (1990).
23 Rieter-Scragg Ltd, *Technical Data Publication SDS 900* (1984).
24 G. Naylor, *Textile Machine*, Rieter-Scragg Ltd., EP0744481 (A1), 27.11.1996.

25 eFK Brochure, www.barmag-oerlikontextile.com/Portaldata/1/Resources/barmag/pdf/barmag_efk_brochure_en.pdf, 2010.

26 G. Naylor, C. Atkinson, *Texturing Yarn*, Rieter-Scragg Ltd,. EP0853150 (A2), 15.7.1998.

27 C. Atkinson, J. Spahlinger *et al.*, Temcooler: Direct Active Yarn Cooling in Draw Texturing, *Chemical Fibers International*, 54, 336, Oct. (2004).

28 Oerlikon-Barmag, *Solutions for the Draw Textured Yarn Production* (2009).

29 U. Wagner, M. Herzburg, High-speed Texturing with highest Flexibility, *Chemical Fibers International*, 54, 332, Oct. (2004).

30 Oerlikon-Barmag, *MPS: As Creative as the Market*, Publication OBA 307 e/7/2007.

31 G. E. Isaacs, C. Atkinson, *Optimisation of Process Performance and Yarn Quality at Current Texturing Speeds, Developments in Texturing*, Shirley Institute Publications S 26 (1980).

32 C. Atkinson, Machine Configurations in High-speed Drawtexturing, Rieter-Scragg Ltd, *Chinatex*, Shanghai, Jun. 10–16 (1984).

33 C. Atkinson, Rieter-Scragg Ltd., *False Twist Method*, GB2133810 (A), 01.08.1984.

34 www.polymer processing.com/polymer/PA66 html.

35 J. Bruske, J. Lünenschloss, Bedeutung der Texturierzonenlänge für den Falschdrahttexturierprozess, *Chemiefasern-Textilindustrie*, 37, 102–106, Feb. (1987).

36 M. J. Denton, *The Throwsters 'from Silk to Synthetics'*, British Throwsters' Association (1994).

37 http://en.wikipedia.org/wki/Glass_transition.

38 www.astm.org/DIGITAL_LIBRARY/STP/PAGES/STP153785.htm.

39 J.W.S. Hearle, L.Hollick and D.K. Wilson, *Yarn Texturing Technology*, Woodhead Publishing Ltd. (2001).

40 M. J. Denton, The Design of Modern False-twist Type Draw-texturing Machinery – An Independent Assessment, *Textile Month*, 57–61, Jun. (1979).

4
Yarn twist application methods

Abstract: This chapter reviews yarn twist application methods. It discusses friction disc twisting technology and materials. It then reviews belt twisting before assessing the improvement of yarn elasticity by torque generation.

Key words: false twist texturing, friction disc twisting, belt twisting.

4.1 Friction disc twisting

The concept of twisting and de-twisting a yarn to satisfy the requirements of false twist texturing is described in Chapter 2, Section 2.2. Friction disc twisting technology was introduced in the early 1970s, resulting in the most significant advance in process speeds experienced in the history of the industry. Friction disc technology remains the most common means of twist insertion used in the industry today.

Twist is applied to the yarn through friction contact with the surface of rotating discs with friction disc technology. Through optimisation of disc diameter, disc surface profile, disc thickness, vertical separation and centre spacing (twist unit spindle separation), twist- and transport-component vectors produce desired twist levels, thread line stabilities, process tensions and textured yarn properties (Fig. 4.1). Friction disc dimensions, surface profiles, material and surface roughness have been the subjects of many development activities in the past, where machine- and component-manufacturers have worked towards optimising disc performance for twist units with different spindle centre spacing.

A change in angle on the friction discs effects the yarn transport and twist components and yarn tension.[1] To minimise yarn abrasive damage due to slippage of the yarn on the friction disc, development activity in the 1970s concentrated on the yarn geometry required to provide a disc-to yarn speed ratio (D/Y), which satisfied 'non-yarn slippage' conditions. Theoretically, it was shown that for non-slip conditions of the yarn on the friction discs:[2]

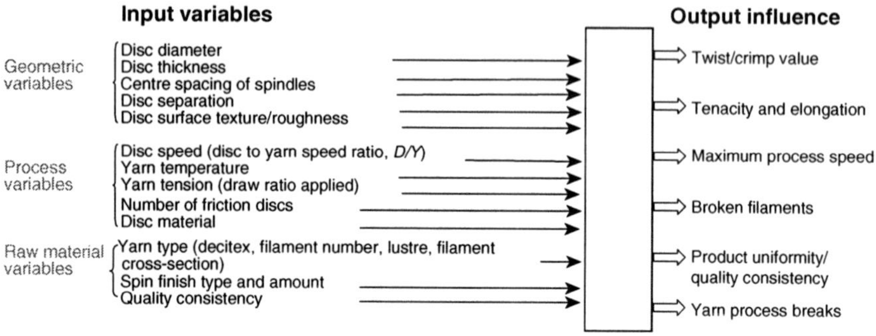

4.1 Twist parameter variables affecting process performance.

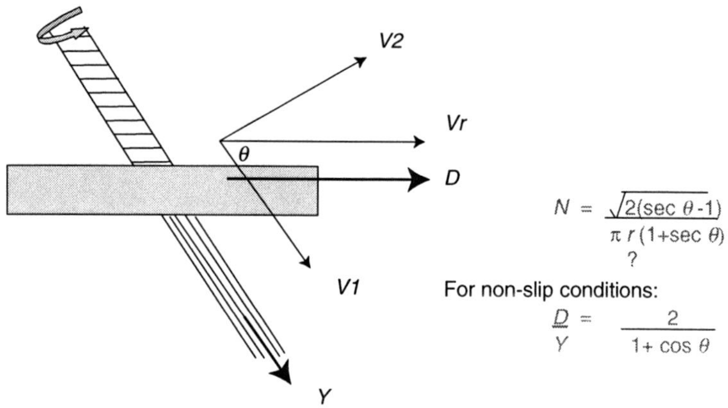

Where:

N	Number of turns of twist per unit untwisted yarn length
ρ	Polymer density
$V2$	Twisted yarn rotational velocity
Vr	Twisted yarn surface velocity
D	Friction disc surface speed
$V1$	Twisted yarn forward velocity
Y	Forward speed of untwisted yarn
r	Equivalent radius of untwisted yarn as parallel filament bundles: $r = \sqrt{\dfrac{\text{decitex}}{\pi.\rho.1000000}}$
θ	Surface helix angle of twisted yarn

4.2 Theoretical considerations for non-slip yarn conditions on friction discs.[2]

$$\frac{D}{Y} = \frac{2}{1 + \mathrm{Cos}\theta} \qquad [4.1]$$

where θ is the surface helix angle of the twisted yarn (Fig. 4.2)

Indeed, the series of twist units, Positorq, marketed by Rieter-Scragg, aimed at achieving low *D/Y* operations through geometrical design of disc diameters, thickness and spindle centre spacings to minimise filament abrasive damage. The Rieter-Scragg Positorq concept also incorporated offset input yarn geometries so that the yarn was introduced to the first friction disc (usually a polished guide disc) at a theoretical angle to minimise slippage.

So, what design parameters of the friction twist unit and friction discs affect the texturing performance? These can be summarised as follows:

- *Centre spacing of the tri-spindle unit.* The maximum centre spacing applied to twist unit designs is largely governed by the space available along the machine to accommodate the twist units. Typical centre spacings are, 36 mm or 37 mm (TEMCO, Barmag-Oerlikon), 34.5 mm (Rieter-Scragg Positorq 2). The centre spacing influences the friction disc diameter required to achieve the desired angle on the discs for a realistic *D/Y* and required twist level for achieving satisfactory crimp levels in the yarn.
- *Friction disc dimensions.* Disc diameter and thickness similarly influence the angle of the yarn on the friction discs and hence twist level and the *D/Y* applied. Disc dimensions were a subject of much development in the past but tend to be standardised today for individual centre spacing.
- *Friction disc surface profile.* Friction discs are available with different surface profile radii and can be purchased in symmetric or asymmetric forms. Profiles have an effect on the yarn friction due to yarn pressure and influence the yarn path over the disc surface, twist level and the optimum settings for *D/Y*.
- *Surface roughness of the friction discs.* Sufficient friction has to be available to provide adequate torque to twist and forward the yarn in accordance with the false-twisting concept. Increasing surface roughness of a hard disc material can slightly increase twist levels, but in many instances, this is offset by an increase in yarn angle on the disc due to the yarn being pulled in the direction of the rotating disc. Optimum *D/Y* settings reduce with increase in disc surface roughness due to increased yarn angles on the discs and increased forwarding action on the yarn. Yarn abrasive damage is also more prominent with increased disc surface roughness, adversely affecting yarn tensile performance and generating powdery polymer deposits on the machine known as 'snow'. On the other hand, too lower surface roughness of hard discs can lead to variation in twisting torque across the machine or gradual friction loss due to increasing contamination levels on the discs. In both cases, product uniformity of the textured yarn is adversely affected, both within- and between processing positions.
- *Number of friction discs.* Increasing the number of friction discs on the twist unit provides more torque to the yarn and increases twist level and

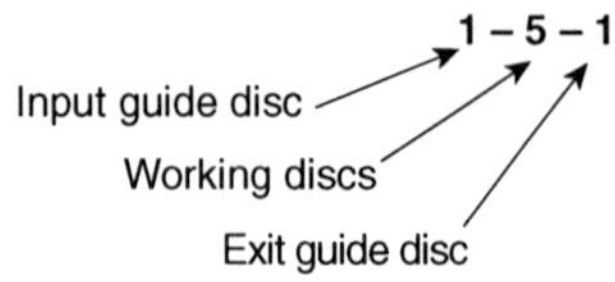

4.3 Friction disc combination description.

crimp value. This is the major parameter affecting twist. However, increasing the yarn contact surface due to disc numbers can adversely affect yarn tensile performance and the maximum process speed attainable due to the onset of process instability (see Section 5.2).

- *Disc material.* This can be categorised as hard discs or soft disc systems. For given geometries, the hard discs rely on surface roughness to generate friction and twist to the yarn. Soft disc systems are predominantly made from polyurethane and these generate friction through 'grip' on the yarn, similar to car tyres on the road! Polyurethane disc hardness is important: the softer the material, the higher the surface to yarn contact friction (see Section 4.2).

Friction disc combinations are described as shown in Fig. 4.3. It is usual to operate with input and exit guide discs; these generate minimum torque to the yarn and simply define the input and exit yarn paths. The selected number of working discs can be variable, depending on twist levels required. The input guide disc is highly polished to avoid filament damage, as the yarn entry angle on this disc tends to be low and the forwarding vectors from the working discs drag the yarn over its surface. De-twisting takes place at the exit disc and similarly its contact surface is highly polished to avoid filament damage.

To maintain good process stability, it is preferable to operate with *D/Y* settings that provide a post- to pre-twist unit yarn tension ratio (*T2/T1*) of between 0.8 to 1.1. If the output tension is too low at the twist unit, there is a danger of short twist lengths slipping through the unit that are manifesting themselves as 'tight spot' real twist faults in the textured yarn. If the output tension is too high, broken filaments, especially on fine filament yarns, can occur during de-twisting and in the post-twist unit thread path. For a given process speed, *D/Y* is set according to the desired *T2/T1* ratio by adjusting the drive speed to the twist units across the machine. The twist units are either driven through wharf contact with a tangentially driven machine-length belt (Fig. 4.4) or more recently by individual twist unit drive motors (Fig. 4.5).

In all cases, the spindle speeds are synchronised through timing belt drives incorporated within the twist units. Yarn threading is carried out with the friction disc drives engaged. Threading is carried out with the use of tines attached to a gate on the twist unit (Fig. 4.4), remote hand-held

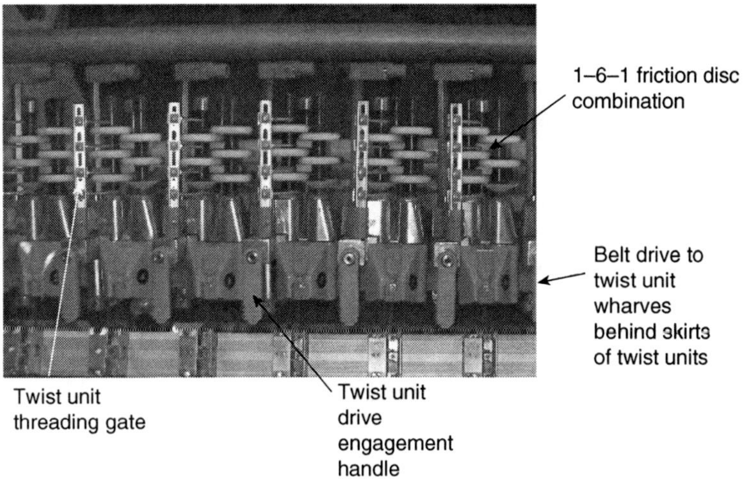

4.4 Barmag Type 8 twist units driven by tangential machine-length belt.

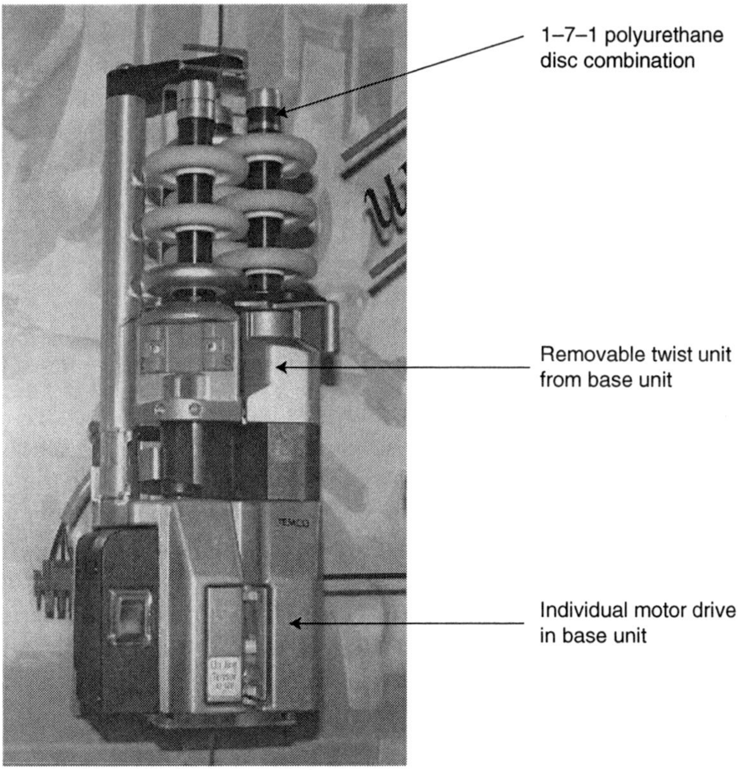

4.5 Individually-motor-driven twist unit.

threading devices that lock onto the twist unit or an open/close twist unit whereby one spindle swings open to allow yarn entry into the disc stack. Of the methods, open/close systems offer the lowest risk of yarn breakage, resultant yarn wraps on the spindles during threading and operator inefficiencies. Individual motor driven twist units:

- Eliminate the risk of twist unit drive slippage on the tangential drive belt, particularly if tangential belt tensions are not set correctly.
- Eliminate the risk of speed variation due to misalignment of twist units with the tangential drive belts.
- Significantly reduce noise emission.

4.2 Friction disc materials

There are mainly three types of friction disc materials used in false-twist texturing:

- Polyurethane
- Ceramic
- Coated
  - Plasma (Ceramic)
  - Nickel–Diamond

In addition, friction disc configurations used on twist units usually comprise low-friction, highly polished input- and output-guide discs, which stabilise the yarn paths at entry and exit to the working disc assembly. These are:

- Entry Guide Discs
  - Ceramic polished
  - Hard chrome polished
  - Plasma coated, polished
- Exit Guide Discs:
  - Ceramic polished
  - Hard chrome polished
  - Stainless steel or ceramic knife-edge

Regarding 'hard' friction disc surfaces, i.e. ceramic, plasma coated and nickel–diamond, for given disc geometries and spindle centre spacing, increasing the surface roughness of the disc tends to reduce the yarn angle θ, as the increase in friction drags the yarn in the surface velocity direction of the friction discs. As a result, the post twist unit yarn tension, $T2$, reduces. This means that for the same disc combination, increasing surface roughness enables the process to run at lower D/Y, i.e. reduced disc surface speed, lowering the yarn transport component and increasing the $T2$ yarn tension.

In these circumstances, there can be both advantages and disadvantages caused by the higher friction. The advantages are:

- Reduced tendency to disc glazing, which is a phenomenon where the effective roughness of the disc is progressively reduced through a combination of polymer and spin finish debris contamination of the disc surface. Disc 'glazing' is complex and its onset is affected by yarn temperature, number of working discs, process tension, disc surface roughness and spin finish type. It is evident through gradual loss of friction, progressively reducing the yarn transport component with increase in *T2* tension, and in advanced stages, reducing twist through inadequate torque provision with increase in tension *T1*.

- Can increase the twisting torque and twist level in the yarn, particularly for higher yarn counts. As a result, yarn bulk levels are increased, but the maximum process speed attainable is reduced because of the onset of 'surging' at lower speeds (see Section 5.2). If this is encountered, it is necessary to reduce the number of working discs, reverting to higher *D/Y* to maintain a *T2/T1* yarn tension ratio, negating the potential low *D/Y* advantages of the rougher surface disc!
- Lower *D/Y* operation, with increased bearing life on the twist unit spindles and lower energy consumption.

The disadvantages are:

- Increased generation of debris due to abrasive damage on the filament surfaces (snow). This comprises typically the order of 60% polymer and 40% residual spin finish and affects environmental working conditions and machine cleanliness. Snow generation also affects the cleaning cycle needs of air-intermingling jets if used, i.e. due to the rate of deposit build up within the jet and resultant loss of intermingling efficiency.
- Abrasive damage caused by the rougher disc on the yarn lowers yarn tensile performance and can increase the tendency to broken filaments.[3]

It is common to specify the roughness of hard friction discs in the surface roughness parameter, *Ra*, which is a universally recognised parameter of roughness. It is the arithmetic mean of the departures Y of the surface profile from the mean line X, and is normally expressed as the mean result from several consecutive sampling lengths, L (Fig. 4.6). For friction disc applications, surface texture *Ra* values are usually specified by the manufacturers within the range 0.85 to 1.1 μm. It must, however, be noted that Ra is not the sole parameter that affects friction disc performance. For example, the two surfaces depicted in Fig. 4.6 have the same *Ra* value, but can be expected to perform quite differently in the false-twist process, particularly in terms of yarn tensile performance.

Surface B will exhibit higher friction on the yarn due to yarn contact on its surface peaks and, for the same *D/Y* and disc number, will provide a higher transport component to the yarn and most likely increased surface filament damage and snow generation. This is demonstrated in results from texturing trials conducted on a PET 78 dtex yarn using solid ceramic 6 mm thick discs, *Ra* = 0.85 μm (Fig. 4.7, Table 4.1):

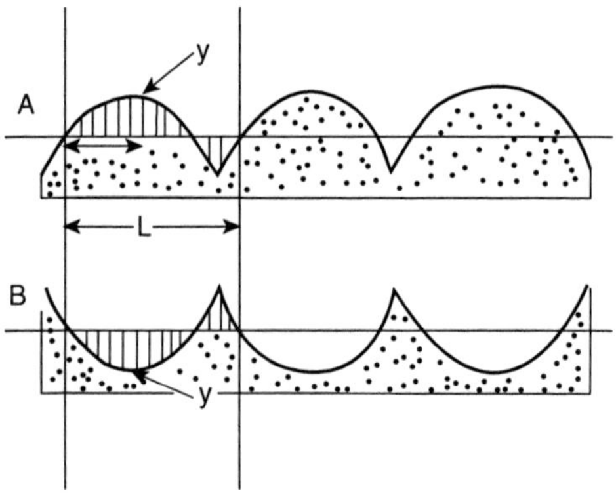

4.6 Two completely different hypothetical surface textures (A and B), which have identical *Ra* roughness values.

Scanning electron micrographs of friction disc surfaces

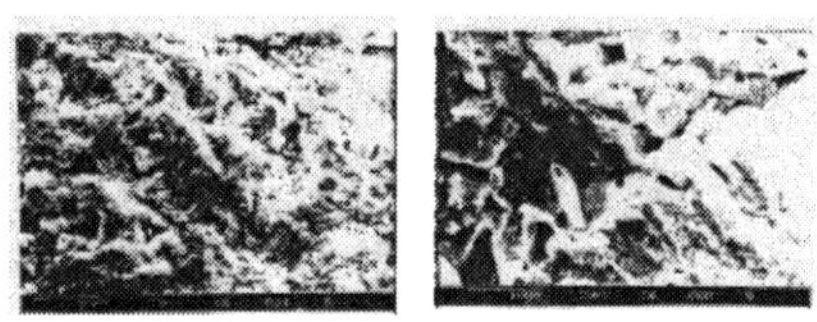

4.7 Two different friction disc surfaces, both with *Ra* 0.85 μm.

Table 4.1 Results from texturing trials conducted on a PET 78 dtex yarn using solid ceramic 6 mm thick discs, *Ra* = 0.85 μm, with the surfaces shown in Figure 4.7

Linear density (dtex)	81.4	80.2
Tenacity (cN/tex)	37.6	35.1
Elongation (%)	25.2	21.7
E (%)	41.7	41.5
K (%)	26.3	26.3
B (%)	87.5	86.1
Tension, *T1* (cN)	19.8	21.0
T2 (cN)	25.2	18.8
T2/T1	1.27	0.9
Machine conditions for above yarns are (Barmag FK6 V80):		
Process speed:	800 m/min	
Disc combination:	1-6-1	
Draw ratio:	1.72	
D/Y:	2.10	
Heater temp.	225°C	
Yarn type:	PET 78 dtex f 34	

There are three types of hard friction discs used commercially in the draw-texturing industry:

- Solid ceramic: most frequently used today.
- Ceramic coated (plasma) on aluminium base disc: commonly used in the 1970s to 1990s.
- Nickel–diamond coated on aluminium base disc: used mainly in the 1970s and 1980s.

The market for plasma friction discs was largely pioneered and dominated by Rieter-Scragg, who manufactured their 'LoSno' brand for their own 'Positorq' twist units, applying special surface finishing methods to maintain disc surface textures within specified tolerances for high torque generation and low filament damage. More recently, a new friction disc type has emerged from Broell, Germany,[4] with a nano-oxide–metal–ceramic coating and lower *Ra* value of 0.5 µm. This disc type is claimed to:

- Achieve yarn tensile properties in line with the high levels achieved with soft polyurethane discs through minimum abrasive damage to the filaments.
- Generate an extremely low amount of snow.
- Be resistant to surface contamination due to its built-in anti-adhesive properties, thus eliminating any tendency to disc surface glazing.

Solid ceramic friction discs are commonly used in the industry today, a leading disc manufacturer being CeramTec, Germany. Solid ceramic discs are heavier than plasma-coated discs and therefore create more stress on the spindle bearings of the twist units, but their surfaces are significantly less susceptible to damage than their plasma-coated counterparts.

The advantages and disadvantages of the three types of hard friction discs used in false twist texturing are listed in Table 4.2. Usually, hard friction discs are cleaned periodically to maintain friction consistency by simple ultrasonic means. Solid ceramic discs, however, have the added advantage that they can be cleaned in high-temperature ovens.

Polyurethane friction discs (Fig. 4.8) provide some real advantages over hard friction disc systems, with the exception of finer count nylon yarn applications. Polyurethane (PUR) is a soft material and historically, discs have been marketed at typically between 80–95 shore hardness. The polyurethane tyre is usually moulded onto a plastic hub, which is designed to precisely retain the tyre during high-speed rotation.

Soft disc systems are, to a large extent, dependent on the reaction force of the yarn (yarn tension) generating surface contact friction. The softer the material used for the disc, the higher the surface-to-yarn friction, and the higher the yarn transport component, but this is often at the expense of the disc life-cycle. Development in PUR-material aimed at improved

Table 4.2 Advantages and disadvantages of hard friction disc systems

	Nickel–diamond coated	Plasma coated		Solid ceramic
		Standard ceramic coated	New variant; nano-oxide-ceramic	
Process issues				
Filament damage; effect on yarn tensile performance and broken filaments	Tends to be high due to surface characteristics	Depends on surface finish	Low	Depends on surface finish
Twist/crimp value of yarn	High	Depends on surface finish	High	Depends on surface finish
Snow generation	High	Low-medium	Low	Low-medium
Disc glazing tendency	Low	Susceptible, but depends on surface finish	None	Susceptible, but depends on surface finish
Maximum process speed	Tends to be lower due to high twist insertion	Depends on surface finish	High due to ability to support high tensions with low tendency to filament damage	Depends on surface finish
Heat transfer	Facilitates cooling due to metallic base	Facilitates cooling due to metallic base	Facilitates cooling due to metallic base	Poor thermal conductor
Maintenance issues				
Cleaning	Ultrasonic, cannot be high-temperature cleaned	Ultrasonic, cannot be high-temperature cleaned	Ultrasonic, cannot be high-temperature cleaned	High-temperature and ultrasonic cleaning possible
Mass (influence on spindle bearings)	Low	Low	Low	High
Susceptibility to surface damage	Yes	Yes	Yes	Very low

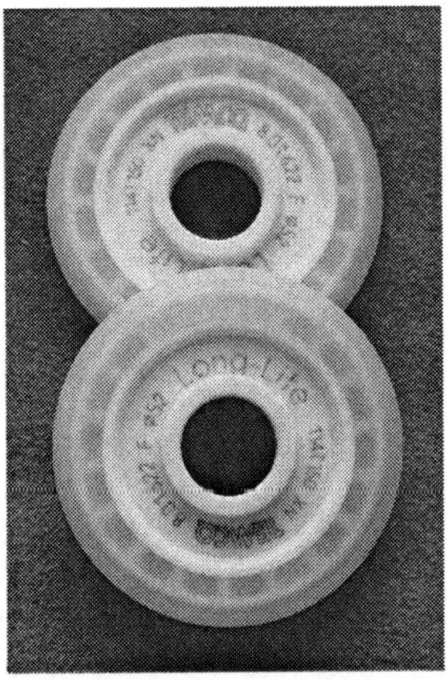

4.8 Example of a polyurethane friction disc.

wear characteristics and chemical resistance, coupled with a wide introduction of spin finishes which are more PUR-material friendly, have ensured that PUR friction discs maintain a firm position in the texturing industry today. Their position has also been strengthened by:

- Growth in texturing requirements for sensitive fine filament yarns (0.2 to 1.0 d.p.f.).
- Increase in clean-process awareness for plant housekeeping and environmental reasons (reduced snow generation from abrasive damage on the filament surface – Fig. 4.9).
- Wide use of on-line process tension monitoring, which is able to detect PUR-disc wear or damage through process tension change.
- Increased use of air intermingling and the associated need for a contamination-free air jet operation.
- The need to maintain high yarn crimp levels at increasing texturing speeds without loss in yarn mechanical performance (reduced filament abrasive damage at the high process tensions necessary to support high processing speeds).

It is well recognised in the polyester false twist texturing industry that PUR friction discs have many advantages, especially concerning yarn properties and their ability to broaden process operating windows, enabling more scope for process flexibility and speed. Their main disadvantage is, however, disc wear, which affects disc life and associated process costs. There are, however, hidden costs incurred with alternative hard disc systems, such as the need for more intensive machine- and component-cleaning due to snow generation, shorter bearing lives on twist units, and the need to clean discs

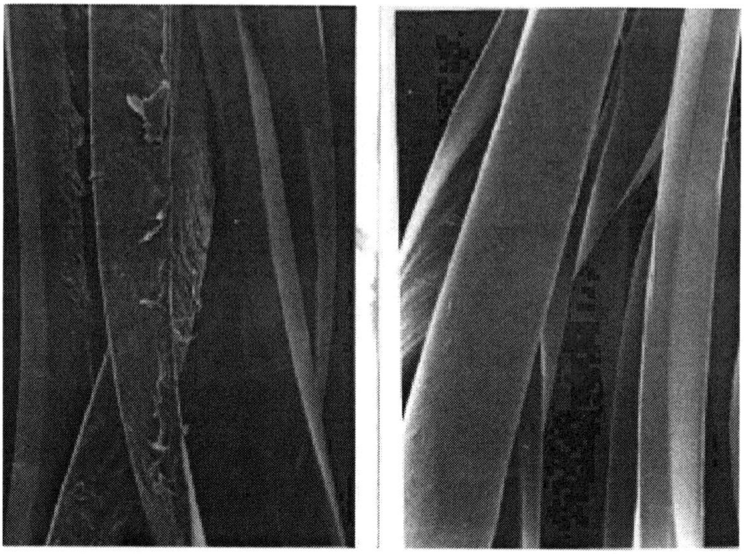

4.9 Scanning electron micrographs of polyester false twist textured 167 dtex f32 Yarn, with Ni–Diamond and PUR disc twisting systems.

where surface glazing is encountered. These factors must be considered when a true cost comparison is made.

PUR disc wear is apparent as:

- Change in surface profile of the disc due to abrasive wear. This tends to be more evident on the output side of the discs and is evident on the first few discs in the stack, which carry out most of the work in twisting the yarn. In disciplined plants, the discs lower down the stack with little or no wear can sometimes be recovered for further use at the end of the disc combination life-cycle.
- Chemical attack on the polyurethane by the residual spin finish on the yarn. This causes accelerated wear through deterioration in mechanical properties and sometimes swelling of the PUR. Polyurethane disc manufacturers usually have tensile testing facilities for measuring tensile performance loss after a pre-determined soak period in spin finishes at elevated temperature in order to determine spin finish compatibility. Today, most spin finishes are designed to be operable with PUR friction discs.
- The onset of 'chatter mark' indentations on the disc surface (Fig. 4.10). This has been reported to be caused by stick–slip of the yarn on the disc surface, stimulated by a complex reaction of longitudinal and traverse

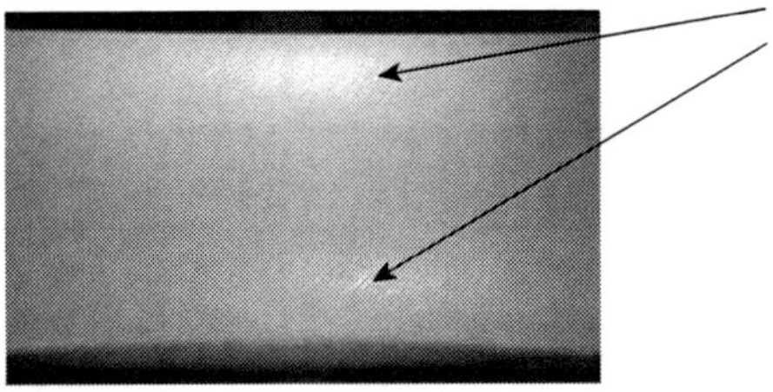

4.10 Evidence of slight chatter marks on PUR friction disc surface.

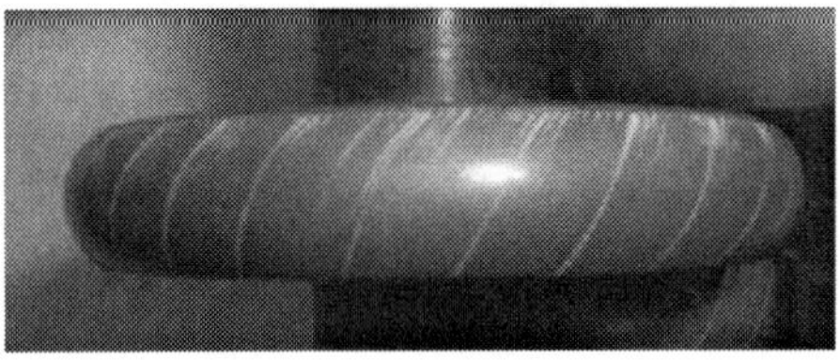

4.11 Periodic snow deposits on hard disc surface.

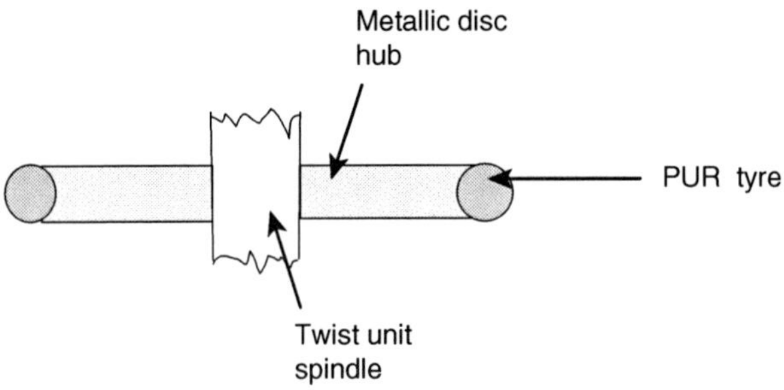

4.12 Removable PUR tyre on metallic hub.

vibrations in the yarn.[5] Yarn tension, process stability, and sometimes eccentricity problems associated with the disc manufacturing quality also have an effect on the chatter mark pattern. The constant impact of the yarn at the chatter points can cause progressive deterioration in mechanical performance of the PUR. This can be accelerated by spin finish attack. The phenomenon of stick–slip is also evident on hard disc systems as periodic patterns of snow deposit (Fig. 4.11).

Attempts to reduce the cost of PUR discs through the manufacture of replacement tyres on metallic hubs did not materialise as a commercial success due to tyre retention problems in high-speed texturing and the inferior eccentricity of the disc (Fig. 4.12). Improvement in PUR material for friction disc application continues to be a focus for development.[6,7] From one such development, by TEMCO, Germany (now part of Oerlikon), it is

Table 4.3 Advantages/disadvantages of PUR friction discs compared to hard disc systems

Advantage	Disadvantage
Better yarn tensile performance and less tendency to broken filaments (minimum filament damage) Little snow generation; reduced cleaning needs, cleaner environment, better air-intermingling security Higher yarn crimp values attainable leading to higher elasticity and better fabric cover Wider process specification flexibility Potentially higher process speeds attainable through support of higher process tensions No tendency to disc surface glazing (exception being fine count PA66 and PA6 yarns) Low mass for extended spindle bearing lives on twist units Reduced energy consumption; tend to operate with lower D/Y settings	Disc life and associated costs of periodic replacement Susceptible to operator damage when removing any yarn wraps from twist unit spindles leading to higher demands on operator disciplines Must have compatible spin finish on feeder yarn

claims that friction disc life can be extended by 30% for the same material hardness.

Because of their mode of operation, PUR discs rely heavily on yarn pressure on their surface to generate friction in order to twist and forward the yarn. They are not generally suitable, therefore, for fine count nylon 66 and nylon 6 applications. For such low yarn tension applications, the yarn tends to slip over the disc surface causing a polishing effect and rapid loss of friction, resulting in twist and yarn quality variation between processing positions. The advantages and disadvantages of PUR discs are summarised in Table 4.3. They are predominantly used in the draw texturing of polyester yarns.

4.3 Influence of friction disc twisting parameters on texturing performance

The twist unit manufacturers set the spindle centre spacing separation. The separation between friction discs is normally between 0.5 mm and 0.7 mm and is precisely set by cylindrical aluminium spacers, which slot onto the spindles between the discs. Today, friction disc diameters tend to be standardised for twist unit centre spacing. In the past, however, because machine manufacturers also designed discs, diameters were often fine-tuned for specific process requirements. Spindle centre spacing, disc diameter (disc

Table 4.4 Texturing 44 dtex f13 nylon 66 with 6 mm thick solid ceramic discs

Disc combination	Measured twist (tpm)
1-6-1	4338
1-5-1	4133
1-4-1	3975

Knitted 44dtex f13 in transmitted light test conditions:

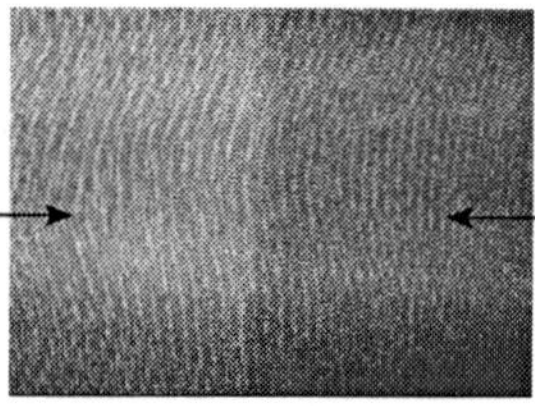

4.13 Effect of crimp value on fabric cover for single-end test knitting.

overlap), disc-thickness and -separation influence both yarn transport and twisting component vectors and the resultant yarn properties.

The ratio between disc surface and yarn axial speed is known as *D/Y*. This ratio has a marked influence on yarn tension ratio *T1/T2* (*T1* = tension before twist unit, *T2* = tension after twist unit). Increasing *D/Y* drags the yarn to an increased angle on the friction disc, thus providing a higher yarn transport component and reducing tension, *T2* (Fig. 4.2). It is often believed in the industry that increasing *D/Y* increases yarn twist but this is not the case; normally *D/Y* has little influence on yarn twist and *T1* tension. There are, however, exceptions, particularly with fine denier yarns, where increasing *D/Y* will increase *T1* tension, reducing twist level, due to a marked increase in yarn angle on the disc, i.e. the twist unit can act as an output feed system to the texturing zone in these circumstances.

Yarn twist is changed by the number of working friction discs on the twist unit. This is demonstrated in Table 4.4.

Increased twist brings about:

- Increased yarn crimp values and higher yarn elasticity.
- Higher cover in the fabric (Fig. 4.13).
- Lower tendency to self-twist (residual yarn torque).

However, increase in number of hard material working discs can:

- Reduce yarn tenacity and break elongation, through increased abrasive damage to the filament surface.

- Increase broken filament level, due to increased yarn contact on the disc surfaces.
- Reduce the maximum process speed attainable, through reduced surge speed (see Section 5.2).

General trends regarding the basic influence of friction disc systems on the draw texturing performance are shown in Table 4.5. These trends are guidelines only and assume parameter changes are within realistic limits.

4.4 Friction disc systems advances

Friction discs have evolved as the leading method of twist application in false twist texturing. Throughout the last two decades, interesting designs have evolved, aimed largely at:

- Improved output efficiencies (reduced machine downtime through ease of threading and yarn wrap removal).
- Ease of disc change for twist direction, disc combination and replacement.
- Increased process speed capability through thread path design within the disc stack.
- Satisfying increasing process flexibility needs (yarn type and crimp values).
- Reduced energy costs.

Of the designs, Positorq 5 (Rieter-Scragg Ltd, 1987) is perhaps the most interesting system in terms of satisfying wider process flexibility needs. On this twist unit, the spindle separation is adjustable via a cam setting, enabling a selection from six settings. This twist unit has the following advantages:

- An ability to change yarn angle on the friction discs, and hence 'dial in' incremental twist changes across the texturing machine.
- A means of changing twist level other than the normal time-consuming disc number change.

The Positorq 5 unit, fitted with 9 mm thickness discs, was introduced to the polyester draw texturing market for increase in twist flexibility. It was aimed at an alternative twist insertion method to the Murata belt twist system (Section 4.5). The twist insertion flexibility of Positorq 5 is demonstrated in Fig. 4.14; it highlights the unit's potential for achieving low twist- and high twist-levels for high and low texturing speeds respectively. The twist unit was often used with a combination of PUR and plasma-coated friction discs.

The true potential of the Positorq 5 concept seems, however, not to have been realised by the market and the unit did not establish a leading position. This is possibly due to:

Table 4.5 Influence of friction disc systems on draw texturing performance

Change		Disc diameter		Disc thickness		Disc separation		Disc roughness	Disc hardness	Number of discs		D/Y	
		Hard disc ↑	PUR ↑	Hard disc ↑	PUR ↑	Hard disc ↑	PUR ↑	Hard disc ↑	PUR ↑	Hard disc ↑	PUR ↑	Hard disc ↑	PUR ↑
Yarn tension	T1	↑	↑	↓	↓	↓	↓	↓	↑	↓	↓	→	→
	T2	↓	↓	↓	↓	↑	↑	↓	↑	↓	↓	↓	↓
	T2/T1	↓	↓	↓	↓	↑	↑	↓	↑	↓	↓	↓	↓
Tenacity		→	→	↑	→	→	→	↓	→	↓	↓	→	→
Elongation		→	→	↑	→	→	→	↓	→	↓	↓	→	→
Crimp value		↓	↓	↑	↑	↑	↑	↑	↓	↑	↑	→	→
Broken filaments		⤴	⤴	↓	↓	→	→	↑	↑	↷	↳	↴	↴
Process stability speed (at same draw ratio)		↑	↑	↓	↓	↓	↓	↓	↑	↓	↓	→	→
Disc wear rate		X	→	X	↓	X	→	X	↓	X	↓	X	↑

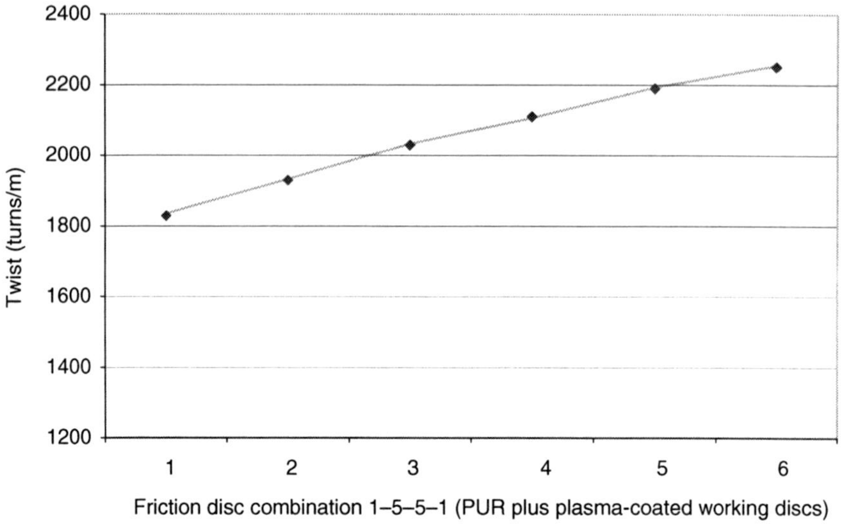

_4.14_ Measured twist levels for polyester 167dtx f34 at each Positorq 5 spindle setting.

- Friction disc replacement and twist change complexity due to top and bottom spindle bearings.
- Mechanical design limitations (potential precision loss of spindle setting through wear and tear).
- Poor accessibility to any yarn wraps on the rear spindle.
- Lack of market understanding regarding the effect of twist level on the maximum process speeds attainable.

Positorq 5, however, was a response to wider process flexibility needs at the time and demonstrated a fundamental understanding concerning the effect of twists level on maximum attainable process speeds.

Other tri-spindle twist unit developments of note largely stemmed from TEMCO GmbH, Germany (now Oerlikon-Barmag) with their FTS525 twist unit. This presented:

- Open/close threading for improved threading efficiency, lower tendency to yarn wraps on spindles during threading and reduced risk of PUR disc damage by the operators.
- Individual motor-driven twist units for better speed consistency and lower environmental impact through energy reduction and noise emission.
- Novel quick S/Z twist direction change capability through simple spindle rotation to avoid machine downtime brought about by friction disc re-stacking.

4.15 Exit knife disc.

TEMCO twist units with their encapsulated bearing systems and the above features are widely used in the industry today.

When friction discs were introduced to false twist texturing, 4 mm thick friction discs were used, often featuring asymmetric surface profiles. Friction disc designs were, at the time, commonly developed by the machine manufacturers in polyurethane or hard materials (plasma or nickel–diamond coatings). Since then, disc thicknesses have progressively increased from 4 mm to 6 mm to 9 mm. At one stage, even 12 mm thick plasma-coated discs were developed for use on Rieter-Scragg Positorq 3 twist units, which incorporated a means of sequentially threading the disc stack using a novel, remote threading device.[8] Discs with 9 mm thickness are commonly used today for yarn counts above 33 dtex, whereas 6 mm discs tend to be preferred for lower count yarns.

The tendency to increased disc thickness has materialised largely for the following reasons:

- On hard disc systems, lower yarn pressure with higher yarn contact surface area enables adequate torque to be generated with lower *Ra* surface roughness values, thus reducing filament abrasive damage.
- On PUR disc systems, longer disc life is obtained due to lower yarn pressure.

In recent years, the use of exit knife discs (Fig. 4.15) has also enabled wider process operating windows to be achieved. These were initially used to reduce tendency to localised, short lengths of real twist ('tight spots') slipping through the twisting disc configuration. They have an added advantage, however, of reducing broken filament levels on finer filament yarns of less than 1 d.p.f. The benefits are realised through localisation of the de-twisting point of the yarn, which otherwise tends to wander on the surface of a standard disc surface radius.

4.5 Belt twisting

Subsequent to the early era of magnetic spindle twisting, a further method of twisting a continuous filament yarn that successfully emerged (other than using friction disc systems) was belt twisting. Belt twisting (Nip Twister) was

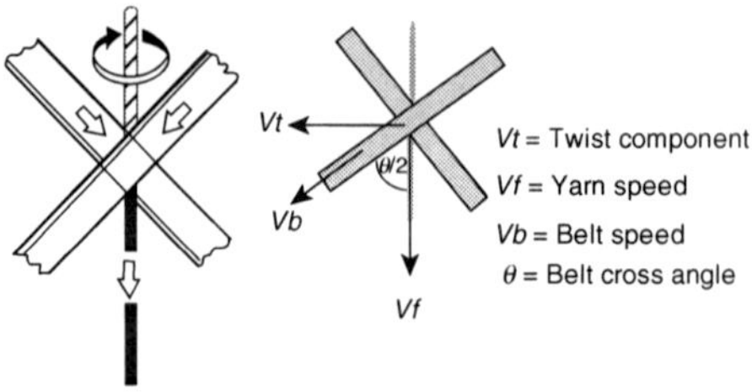

Nip Twister on TMT AFT-12 Machine

4.16 Nip twister system. (Illustrations supplied by TMT Machinery, Inc., Japan.)

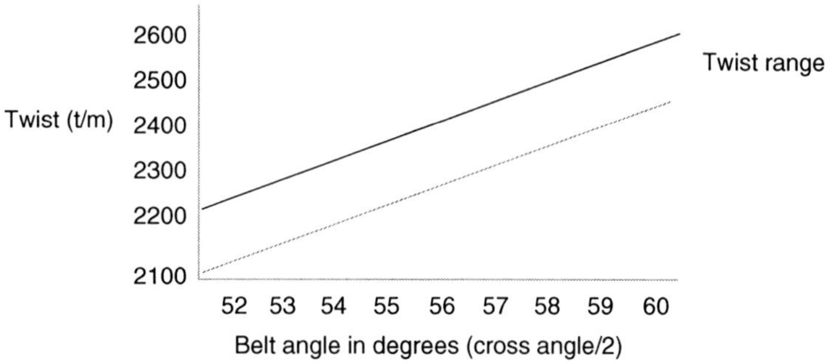

4.17 Typical twist level with change in belt angle for polyester 167 dtex f34.

introduced commercially to false twist texturing in approximately 1978 by Murata, Japan. The system sandwiches the yarn between narrow rotating belts (Fig. 4.16).

In friction disc systems, the yarn contact angle on the discs is controlled by the disc diameter, thickness, surface profile, yarn tension and a friction. In belt twisting, however, an external force is applied to the yarn by the sandwiching effect of the belts and there is a positive grip mode of operation, providing yarn transport and twisting components through the function of speed ratio (belt-to-yarn) and belt crossing angle.[9] The belts are manufactured from rubber and, like PUR friction discs, inflict little surface damage to the yarn.

With the Nip Twister concept, twist levels are changed by changing the belt angles (Fig. 4.17) and to maintain a desired yarn tension ratio *T2/T1*,

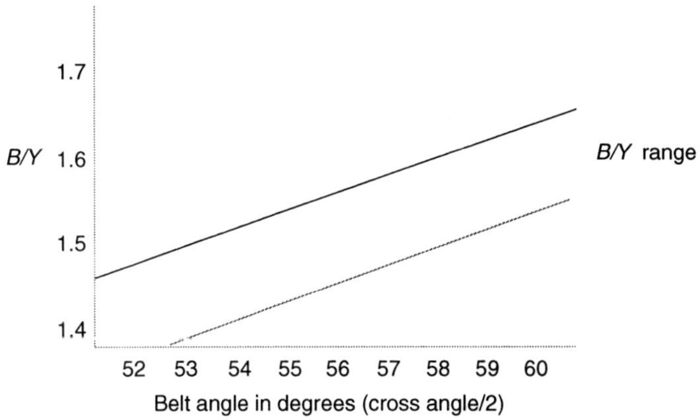

4.18 Typical *B/Y* ratios to maintain constant tension ratio 0.8 for change in belt angle.

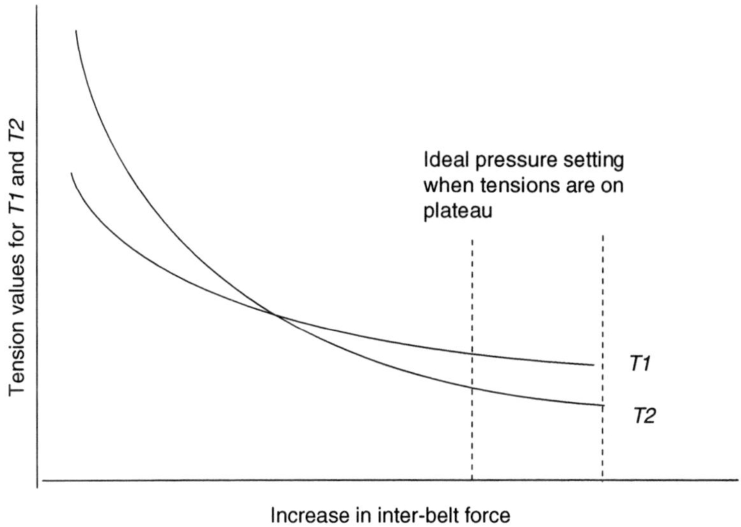

4.19 Influence of inter-belt pressure on process tension.

the *B/Y* (belt-to-yarn speed ratio) must be changed for change in belt angle (Fig. 4.18). Inter-belt force is also a critical parameter; it must enable sufficient friction force to twist and forward the yarn. When too high, belt wear is accelerated through increased inter-belt contact force and friction. Inter-belt force is a critical parameter to maintain sufficient twisting torque and consistent twist levels across the machine (Fig. 4.19). The optimum inter-belt pressure applied increases with yarn count. The pressure is maintained across the TMT machines pneumatically and there is an electronic feedback system from on-line *T2* tension measurements to adjust individual twist unit pressures to minimise *T2* variation across the machine.

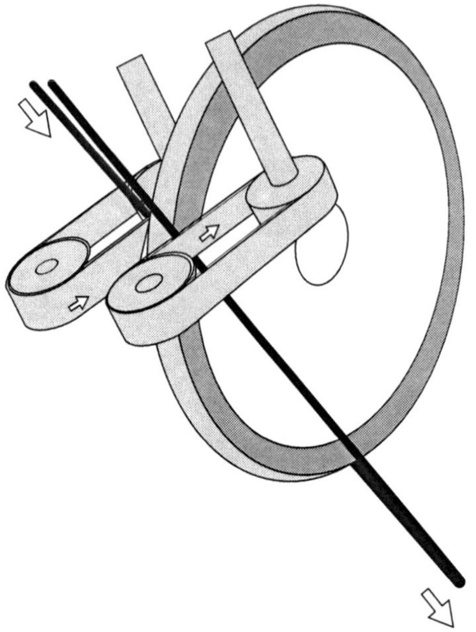

4.20 Nip twister system for two-fold texturing. (Illustration supplied by TMT Machinery, Inc., Japan.)

The Nip Twister system is most suitable for the draw texturing of higher yarn linear densities, e.g. 78 dtex to 330 dtex. Low yarn tensions and twisted yarn diameters associated with the finer yarn counts are more difficult to control at the belt interface, and inter-belt contact increase can be assumed to be more problematic. A TMT system is also available utilising disc and belt for two-fold operations on single texturing positions (Fig. 4.20)

Because belt twisting utilises 'soft' material for the twisting belts, textile performance can be considered comparable with PUR friction disc systems but there remain some advantages and disadvantages when comparing the two systems (Table 4.6). The real advantage of the belt twisting system is being able to dial in required twist levels to attain the highest speed for a desired crimp value. Moreover, yarn tensile properties are excellent and the system is favourable for supporting high yarn tensions (draw ratio) for high-speed operations. Belt condition, however, is critical, especially for fine filament yarns, where damage or wear to the yarn exit side of the belts can cause broken filament formation during de-twisting.

Simultaneously to the development of the Nip Twist belt system, Barmag in the 1980s introduced 'Ringtex', which operated on the same principle of sandwiching the yarn between two contact surfaces. Here, Barmag adopted the principle of clamping the yarn between two rotating discs, the interface pressure between which was pneumatically controlled.[10,11] Ringtex was,

Table 4.6 Comparison of PUR friction disc and belt twisting systems

Issue	PUR friction disc system	Nip twister
Twist flexibility	Incremental change via disc number and type	Fully flexible through belt angle change
Yarn count range	Wide range applicable through disc type selection and number	More suitable for higher counts, ≥78 dtex
Crimp values	Satisfactory range	Higher crimp values can be achieved with high belt angles
Filament surface abrasive damage	Low	Low
Product quality consistency across machine	Good within life span of discs	More plant disciplines required regarding precision of yarn path at belt intersection and consistency of inter-belt pressure
Threading	Risk of spindle wraps and PUR disc damage (Lower risk with open-close twist units)	Good
Disc/belt change	Slow and work intensive	Quick
Maximum process speed	Simply dependent on twist, tension, yarn count and texturing zone profile	Can fine tune twist to satisfy higher speeds but with crimp value loss. Also dependent on twist, tension, yarn count and texturing zone profile
Cost elements	Disc replacement cost high but dependent on life cycle	Replacement of belts cheaper than set of friction discs with similar life cycles
	No data on comparative power consumption	
Retrofit	Twist units can usually be retrofitted to wide machine range	Limited to TMT (Murata) machines

however, not widely adopted by the industry, possibly due to disc interface pressure in relation to disc rigidity issues. Friction disc systems remained the preferred choice on Barmag machines.

4 6 Improving yarn elasticity by torque generation (self twist)

In false twist texturing, yarn elasticity and bulk volume are usually generated through the heat setting of filament crimp in the yarn. There are, however, exceptions to this where elasticity is required in the textile garment with a high transparency and sheer. This was especially the case in the 1980s to 1990s where there was large demand for such effects in fine denier nylon ladies' hosiery. More recently, however, there has been a tendency to achieve the effect through the conventional covering of elastanes and fine count flat yarns to produce the elasticity and sheer.

Concerning draw textured yarns in sheer hosiery applications:

- Sheer and elasticity can be achieved with yarns that tend to self twist through torque liveliness.
- Sheer effect yarns tend to have low filament numbers, e.g. 17 dtex f3 or monofilament.

Typical draw texturing twist levels of around 7000 tpm for a 17 dtex nylon yarn result in too high a crimp value and opacity for the desired sheer effect. Reducing the number of friction discs on a twist unit is a means of reducing twist to the desired ca 2000 tpm level for the sheer effect. This can, however, be achieved with normal disc spacing only at extremely low *D/Y*, resulting in high *T2* tension and a high tendency to yarn damage as a result of the generation of insufficient transport component to the yarn. To operate in this mode with friction discs demands much discipline, and position-to-position consistency of twist can be poor, leading to irregularities in the finished fabric.

Magnetic spindles have been commonly used for torque yarn production. Here, exact twist levels can be applied, but process speeds are limited. Realising the market potential for torque yarns in the 1980s, both Barmag and Rieter-Scragg introduced novel friction twisting variants. The Rieter-Scragg Hitorq and the Barmag Torquemaster both operate with low angles of yarn wrap and hence twist helix angle on their friction surfaces.[12,13] The Hitorq (Fig. 4.21) comprises a plasma-coated cylinder, which can be interchanged with friction discs on a Positorq twist unit spindle. It is designed in such a way that the yarn experiences a helical path over a cylindrical surface. Twist levels are set by an incremental setting of the output ceramic guide. To generate sufficient torque via high yarn pressure on the cylinder and friction, *D/Y* ratios are low (typically less than 1.0), with high *T2* tensions. As a result, the principle operates in a high yarn slippage mode. Typical twist level and torque level ranges of Hitorq are shown in Figures 4.22 and 4.23 respectively. The Barmag Torquemaster operated on the same principle as Hitorq. Their twist unit was, however, retrofitable on a bay length basis, i.e. per 12 positions on the machine (Fig. 4.24). The influence of parameters on both twist and torque for the Torquemaster are shown in Fig. 4.25.[13]

The increased use of elastanes in four-feed fine gauge hosiery knitting since the 1980s for comfort and fit in ladies' hosiery through yarn covering

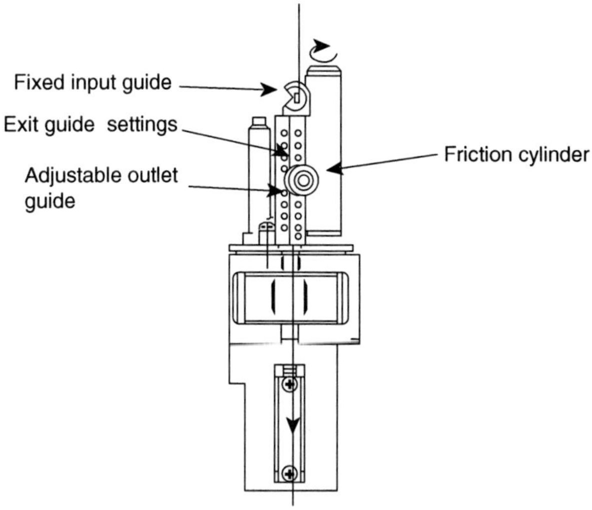

4.21 Rieter-Scragg Hitorq twist unit.[12]

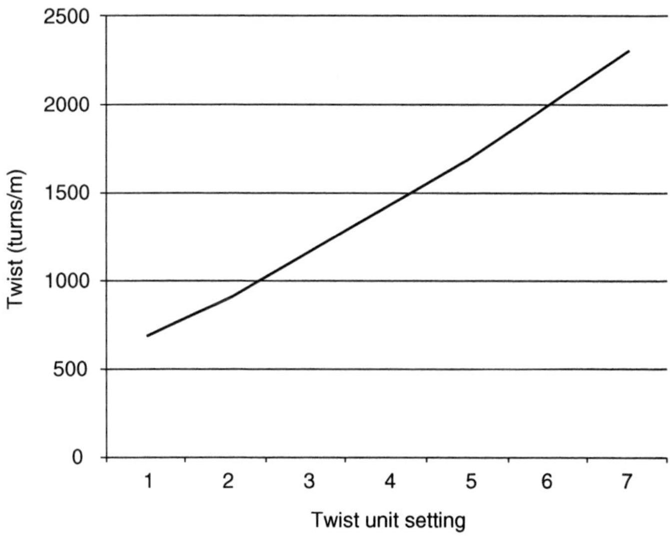

4.22 Twist levels for Hitorq settings, nylon 66 17 dtex f3.

or plating directly in the knitting process, resulted in the increased use of drawn, non-textured yarn to be used for sheer-effect hosiery. Consequently, the amount of torque-textured yarns produced over the last decade significantly has reduced and the torque yarn process is no longer prevalent in the industry.

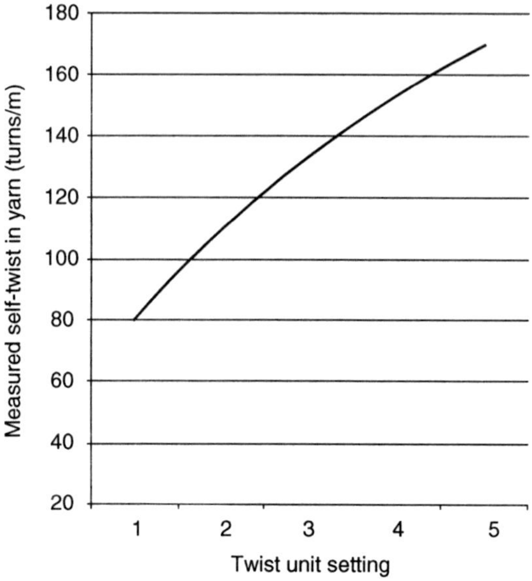

4.23 Torque levels over a range of settings, nylon 66 17 dtex f3.

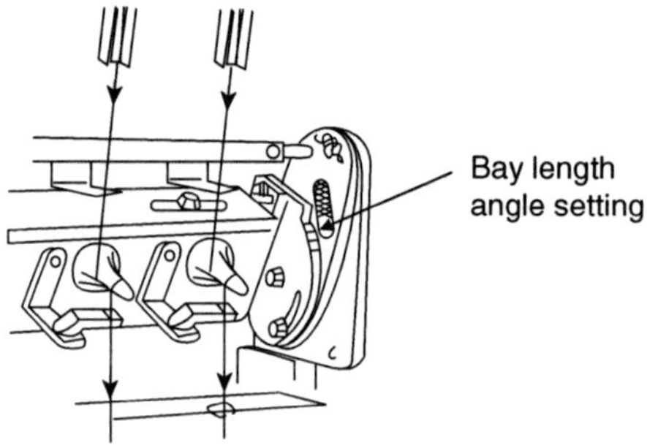

4.24 Barmag Torquemaster.[13] (Illustration © Oerlikon-Barmag.)

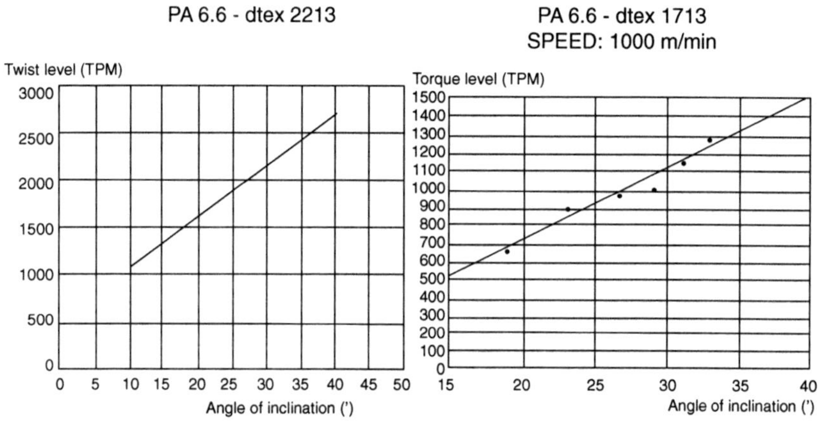

4.25 Influence of the Torquemaster parameters on twist- and torque-level.[13] (Illustration © Oerlikon-Barmag.)

4.7 References

1 H. Weinsdörfer, Aspekte der Friktionstexturierung, *Chemiefasern Textil-Industrie*, 540–551, Jun. (1978).

2 W. J. Morris, *Journal of the Textile Institute*, Vol. 66, 3, 123–128, March (1975).

3 Chr. Richter, R. Hesse *et al.*, Raster-electronen-mikroskopische Untersuchung der Oberflächenveränderung an Filamentgarnen durch Friktionsscheiben, *Chemiefasern-Textilindustrie*, 37, 998–1002, Oct. (1987).

4 E. Sonntag, J. Bolze *et al.*, conTEX: New Generation of Friction Discs, *Man-made Fiber Year Book*, 78, Chemical Fibers International (2006).

5 J. J. Thwaites, *Journal of the Textile Institute*, 75, 76 (1984).

6 C. Atkinson, S. Müller-Probandt, Opportunities for Further Optimisation of the Primary Zone in False Twist Texturing, *Man-made Fiber Year Book 2006*, Chemical Fibers International.

7 A. Weber, Heberlein Fiber Technology Inc., Latest Developments in Friction Texturing, *Chemical Fibers International*, 54, 328, Oct. (2004).

8 C. Lawrence, P. M. Wilkinson *et al.*, Rieter-Scragg Ltd., US4510744 (A), 04.16.1985.

9 Wel Li, Tae Jin Kang *et al.*, *Mechanics of High Speed Texturing, Part III: Experimental Studies of Belt and Ring Twisting*, Textile Research Institute, 719–725, Dec. (1988).

10 K. Bauer, *Chemiefasern Textilindustrie*, 32, 702 (E70) (1982).

11 K. Bauer, *Man-made Fiber Year Book*, Chemical Fibers International, 96, (1987).

12 *Technical Literature, Hitorq*, Rieter-Scragg Ltd, 1981.

13 *Technical Literature, Torquemaster*, Barmag, 1983.

5

Process performance issues in yarn texturing

Abstract: This chapter reviews process performance issues in yarn texturing. Parameters discussed include: yarn contact surfaces, the problem of 'surging', creel design, package building, oil application systems and automatic doffing systems.

Key words: false twist texturing, yarn contact surfaces, surging, creel design, automatic doffing systems.

5.1 Friction and surface texture of yarn contact surfaces

This section discusses yarn contact surfaces. Yarn contact surfaces within the draw texturing process should generate low yarn contact friction to minimise filament damage and ideally not inhibit twist transfer in the texturing zone. Moreover, yarn tension profiles within the sub-zones of the texturing process, i.e. from the creel of raw material POY through to winding of the packages, are influenced by friction and are consideration points for assurance that operating windows for process specifications are not restricted. Frictional properties in draw texturing are indeed demanding and are also affected by inconsistencies in thread path settings. Some examples are discussed below.

The creel thread paths comprise filament yarns, usually POY, which experience tension fluctuations due to their wind reversal points during 'over-end' off-winding. Also, as the POY package diameters become smaller during off-wind, the rotational speed of the off-winding yarn becomes higher and there is a progressive change in balloon characteristics. As both mean tensions and peak tensions rise with reduced package diameter, tensions increase cumulatively immediately prior to the input feed to the texturing zone due to each surface contact and angle of wrap in line with the capstan equation:

$$\frac{tb}{ta} = e^{\mu\theta} \qquad [5.1]$$

where: tb = output tension
ta = input tension
μ = coefficient of friction
θ is the angle of contact (radians) on a cylindrical surface.

Yarn tensions in the creel increase with process speed. High angles of creel wrap and/or high coefficient of friction due to the selection of inferior ceramic guide surfaces can lead to high peak tensions prior to the input feed, which can be transmitted through the input nip feed (slippage) or, if high enough, lead to intermittent cold drawing of the yarn. In such instances, dye uptake consistency on the textured yarn can show as defects in the resultant textile fabric.

Spin finish applied to the POY feeder yarn (typically in the order of 0.3–0.5% by weight) in spinning, comprises lubricant, antistatic compound and emulsifier.[1] A percentage of the finish is volatilised during passage of the yarn through the texturing heater. Lubricant fuming problems affecting air contamination and condensation on cooler machine parts should be minimised as they can change yarn frictional characteristics, especially on cooling tracks. Moreover, any resinous or carbonaceous deposit build-up from the spin finish on heater tracks, as a result of degradation or condensation, can adversely affect textured yarn quality and process performance. In general, spin finishes should ensure low yarn-to-ceramic friction, together with low filament-to-filament friction. High yarn friction on contact surfaces in the pre- and post-twist unit thread paths can lead to snow generation, filament surface damage and broken filaments. High inter-filament friction can cause poor twist migration of the filaments at the heater entry, leading to loopy crimp characteristics, lower crimp values and sometimes a tendency to filament snagging in both textured yarn package off-wind and in downstream fabric manufacture.

Contact surfaces (heater track, ceramic guides, cool tracks), and yarn reaction forces due to various machine profiles, ceramic guide types and diameters vary in the process for machine type and manufacturer. Surface finishes must be especially carefully selected for thread path zones according to whether the yarn is in a twisted of flat state.

During texturing start-up, high-speed yarn contact is experienced on some texturing zone thread path surfaces when the yarn is in a flat state. Subsequently, these surfaces are subjected to twisted yarn during processing. Optimum surface finishes, however, are not the same for both yarn states; highly polished guide surfaces necessary for low friction on the twisted yarn cause high frictional forces on flat yarn, leading to operator difficulties, yarn breakage and poor start-up efficiencies of machine positions.

It is common to apply lubricant (coning oil) to the textured yarn to obtain low friction levels in downstream processing. The properties of coning oils are important in facilitating textured yarn package off-wind, together with low static electricity and friction in downstream processing. Typically around 1.5–2.5% by weight of coning oil is applied to textured yarns. As texturing speeds have increased, so have package wind traverse speeds and, as a result, coning oils have been progressively developed to provide 'low sling' properties to reduce air contamination and slippery under-foot conditions for the operators.

Machine components that have contact with the yarn must be set consistently across the machine, particularly in the texturing zone, so that angles of yarn wrap and surface contact forces remain the same. Cool track settings are particularly often neglected in draw texturing plants, leading to inter-position inconsistency in crimp values, poor yarn tensile properties, broken filaments and rapid wear of the cool track surface (Figures 5.1 and 5.2).

In false twist texturing, the mode of friction can be considered to be either boundary or hydrodynamic, depending on the dynamic situation of the yarn

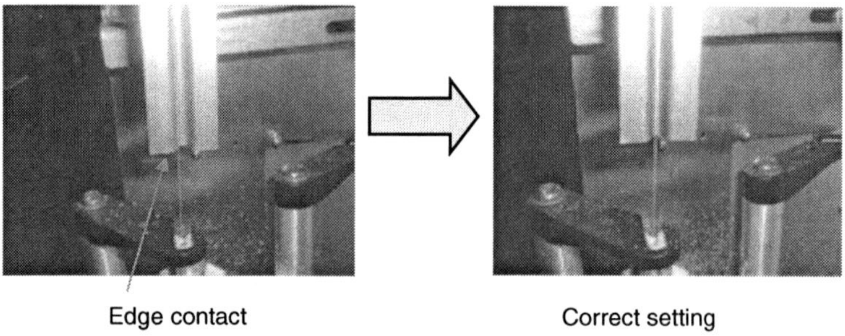

5.1 Cool track misalignment with yarn running on exit edge of track.

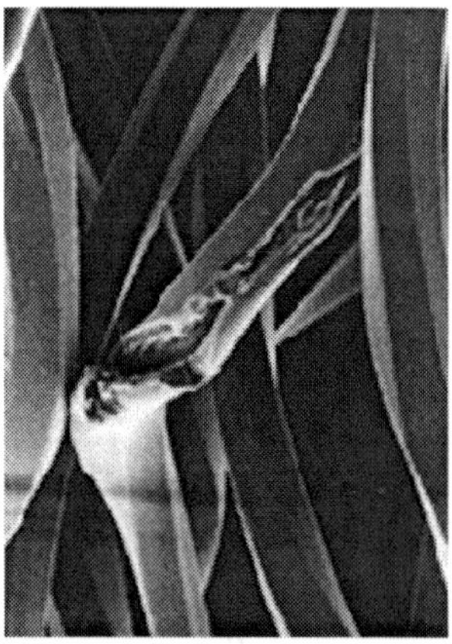

5.2 Filament tear due to cool track edge contact.

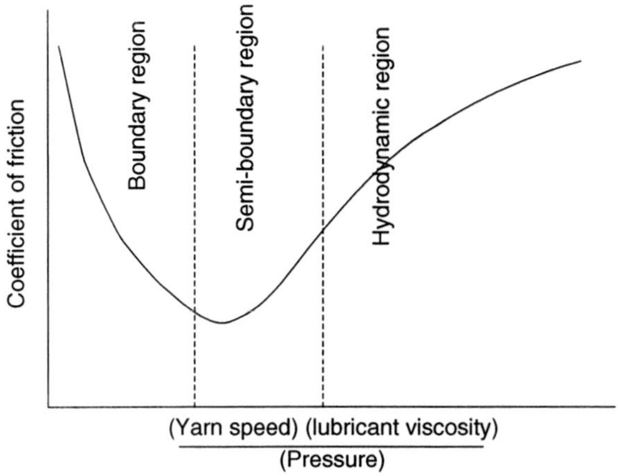

5.3 Modes of friction with respect to yarn speed, lubricant viscosity and yarn contact pressure.[2]

and its contact surface (Fig. 5.3). The mode of friction is largely influenced by speed, lubricant viscosity and yarn pressure. For hydrodynamic friction conditions, the yarn can be assumed to be separated by a lubricant film over which it slides. For boundary friction, the yarn can be assumed to have direct contact with the surface. In general, hydrodynamic conditions can be expected to apply at high processing speeds, provided the surface roughness peaks do not penetrate the lubricant layer.[2] In the creel thread path, yarn tensions and reaction forces (pressure) on guides tend to be relatively low. Lubricant viscosity (low temperature) can also be assumed to be at a relatively high level. Rougher guide surfaces tend be used to reduce the coefficient of friction and hence lower the pre-input feed yarn tension in accordance with a boundary mode trend.

For twisted yarn, however, at the heater exit, lubricant condensate tends to form on guide surfaces; residual lubricant viscosity can be expected to be low (high temperature), yarn pressures due to individual filament contact is relatively high, and yarn surface speed is higher due to the draw ratio and twisting rate. Highly polished guide surfaces improve twist transfer to the heater, suggesting a hydrodynamic mode of operation. Rougher guide surfaces create abrasive damage to the filaments, leading to lower yarn tensile performance, broken filament formation and the generation of snow. This is particularly an issue on folded thread paths in the texturing zone, e.g. Barmag M or Rieter-Scragg B profile machines. Twist transfer over guide surfaces was indeed a subject of much investigation during the folded thread path era.[3–5]

Textile guides are largely manufactured from aluminium oxide or titania (titanium oxide). Titania guides tend to have a finer grain structure and

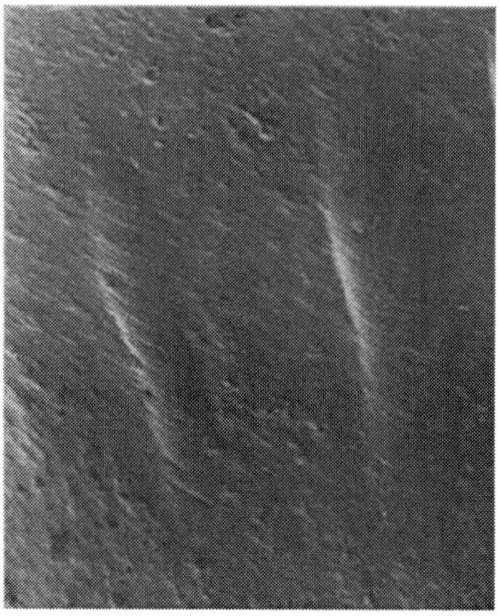

5.4 Evidence of abrasive wear from twisted yarns running over titania ceramic guide.

bring about lower tendency to filament abrasive damage (i.e. the grains are less likely to penetrate protective lubricant films). They are particularly gentler on the yarn in the turn-round region between heater and cool track for folded texturing zones, where yarn reaction forces are high on the guides. They are, however, susceptible to abrasive wear due to their low material hardness (Fig. 5.4). Aluminium oxide has tended to replace titania guides in specific locations on the thread path. Improvement in ceramic surface polishing techniques and developments in material composition have ensured its use as a suitable replacement.

5.2 Process instability: tension transients and speed limitations due to 'surging'

There has been a significant increase in texturing process speeds since the advent of friction disc twisting technology. To a large extent, this increase is attributable to:

- Improved POY feeder yarn quality.
- Machine profile advances (heat–cool zone geometries and reduced length).
- Continuous improvement in machine and component mechanical designs.

There still remains, however, a key factor that limits further breakthrough in further significant speed advance in false twist texturing. This limitation is known as 'surging'.

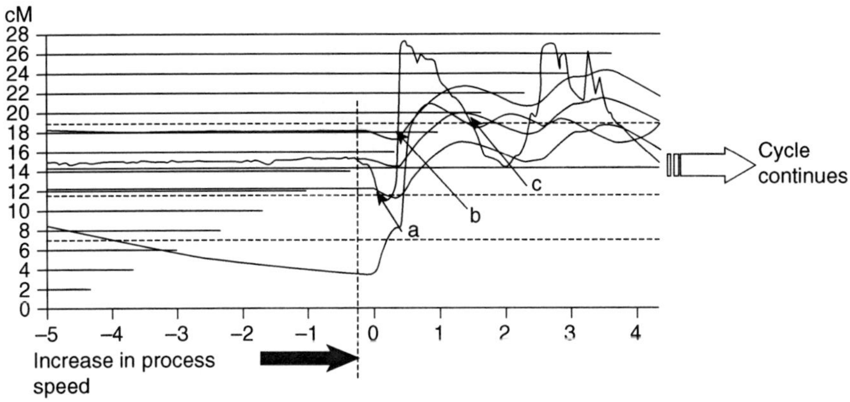

5.5 Online process tension, T_2 surge characteristic.

5.2.1 Continuous surging

For all draw texturing processes, where the process speed is progressively increased, a speed is eventually reached where a periodic 'Surging' in yarn tension begins. Surging manifests itself in the following sequence of events (Fig. 5.5):

(i) A speed is reached where a sudden increase in resistance to twisting occurs in the thread line, the force of which exceeds momentarily the twisting torque provided by the twist insertion system. As a result the twist slips through the twist unit (a).
(ii) Due to the immediate loss of twist level in the heat–cool zone, the yarn tension rises instantly (b).
(ii) The tension level reduces as the twist insertion system gradually regains control on the yarn, imparting twist again to the running thread line (c).
(iv) The twist level rises and the tension (*T1*) reduces to the original level at which the onset of the tension surging occurred; the whole surging cycle is initiated again.
(v) The surging cycle (i to iv above) repeats on a continuous basis, unless conditions are met that move the texturing process below the threshold levels at which surging occurs (or clearly, if the yarn breaks).

In a test circular knit fabric, surging was evident, both in yarn structure and dye uptake variations (Fig. 5.6).

Surging has been a constant focus of attention for machine manufacturers for many years, but to date the author is unaware of publications defining its true cause. It is, however, now widely known that the following process conditions on a given thread path have the most significant effect on the process speed at which surging occurs:

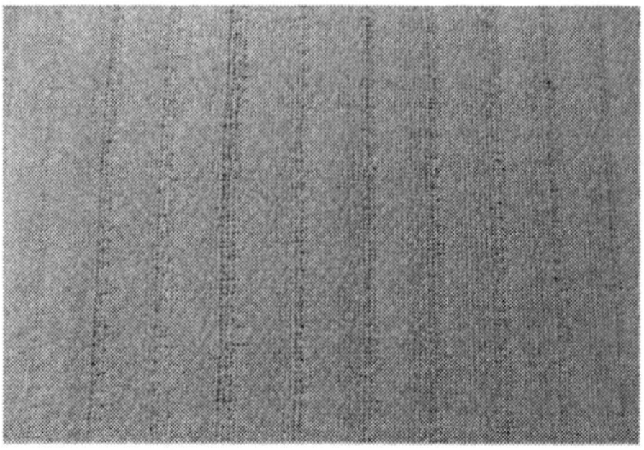

5.6 Evidence of surging in dyed test knitting.

- Twist increase ▲ = surge speed decrease ▼.
- Tension (draw ratio) increase ▲ = Surge speed increase ▲.

Both texturing zone geometry and design, and indeed raw material properties, can also influence surge speed.

Machine factors include:

- Texturing zone length (length ▼ = surge speed ▲), which will be explained later.
- Yarn slippage at the feed system (can cause tension and twist fluctuations, which reduce surge speed).
- Any tendency to yarn ballooning in the texturing zone (causes twist and tension fluctuations, which reduce surge speed).
- Machine profile (for a given draw ratio, folded thread paths bring about higher surge speeds).

Raw material factors include:

- Spin finish type and, to a lesser degree, spin finish level consistency (affects filament migration, twist level and process tension).
- Draw force inconsistency or elongation increase. (Tension loss, whether long-term drift or momentary change in draw texturing, affects surge speed.)
- Yarn levelness inconsistency (Uster U%).

Clearly, continuous tension surging is completely unacceptable in draw texturing because the yarn variation, which comprises periodic textured, virtually flat and twisted yarn lengths, is evident in the downstream fabric made from the yarn.

5.2.2 Single surge events (tension transients)

On-line tension measuring in draw texturing often highlights short-term and random tension disturbances (Fig. 5.7). It is not commonly recognised

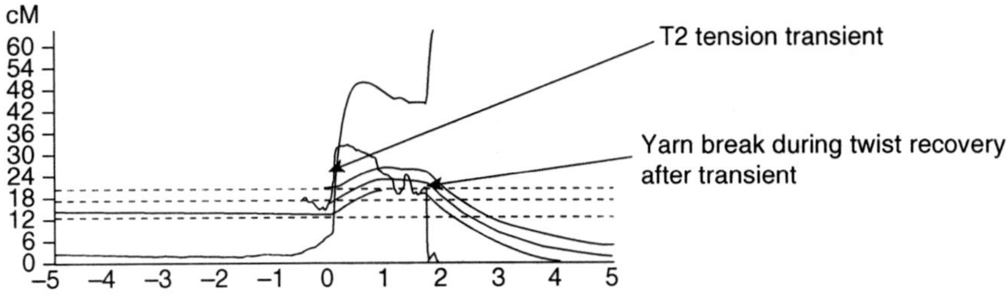

5.7 Example of a tension transient during online tension monitoring.

that these faults are linked to surging and its inherent causes, as will be explained later. Such short-term faults tend to be more common in nylon 66 texturing than in polyester or nylon 6 texturing. In some end-uses for polyester yarns, these faults are particularly undesirable. For example:

- They are evident as long flashes in woven fabric (particularly in warp threads) and are often not detectable until after fabric dyeing and finishing. Because of the added value in process costs, downstream fault detection can be costly.
- In critical applications, such as automotive velour (circular knit), they are evident as stripes (optical and dye shade) in the fabric pile, and are totally unacceptable.

In fine-gauge hosiery end-use for textured yarn, their infrequent occurrence is not as critical as in fabric, with minimum hosiery leg rejection in knitting and at garment final inspection. They are, however, a major source of yarn breaks in the texturing process for high-speed texturing of fine decitex nylon 66 yarn, and can be a significant contribution to process inefficiency and textured yarn package size rejections, with associated costs. Any variation in thickness of the continuous filament yarn entering the twisting zone in texturing can cause single surge events. Variations from POY raw material include:

- broken filaments,
- individual filament draws (snagging and filament separation on the POY package),
- filament loops,
- tail transfer splices/knots for POY package transfer in creel,
- handling damage to filaments, and
- short-term filament diameter variations (gels in nylon 66 extrusion) (Fig. 5.8).

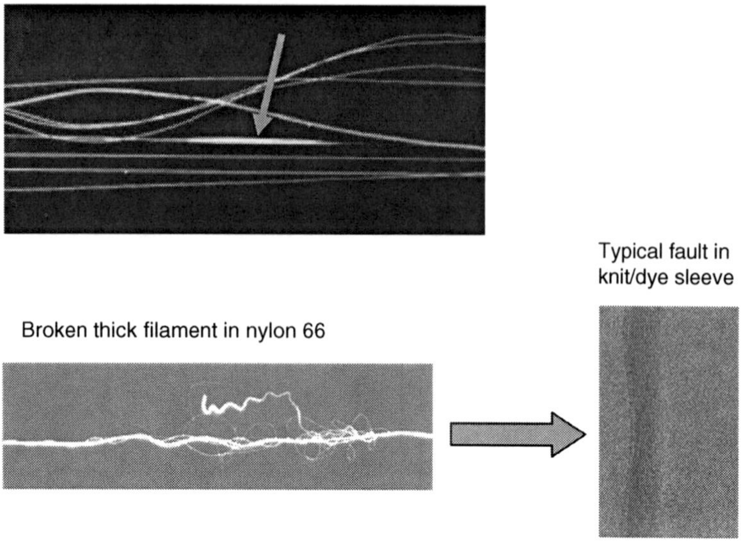

5.8 Gel in nylon 66 POY forming short, thick filament.

Variations from the texturing process include:

- broken filaments (pre-twisting),
- package creel mis-alignment (can cause filament separation during POY off-wind, and
- yarn feed slippage in the texturing zone.

In low twisting torque situations (low decitex yarns, smooth or glazed disc surfaces of friction discs, low number of friction discs), twist may not be recovered after single surge occurrences. Such instances are particularly problematic with respect to yarn quality; without on-line tension measuring in the texturing process, they are not detectable, resulting in long durations of flat, un-textured yarn within textured packages. Similarly, the yarn can occasionally be ejected from the twist stack during tension transient events (Fig. 5.9) and if there is no yarn break, the process will continue running at high *T1* and *T2* tensions, producing yarn with extremely low twist- and crimp-values until detected and rectified through re-threading and re-starting the processing position.

5.2.3 Causes of surging

In the texturing industry, it is often wrongly thought that increasing the friction and grip on the yarn through an increased disc numbers or increased disc surface roughness increases the surge speed limit of the process. In practice, the opposite occurs; with increase twisting torque through an

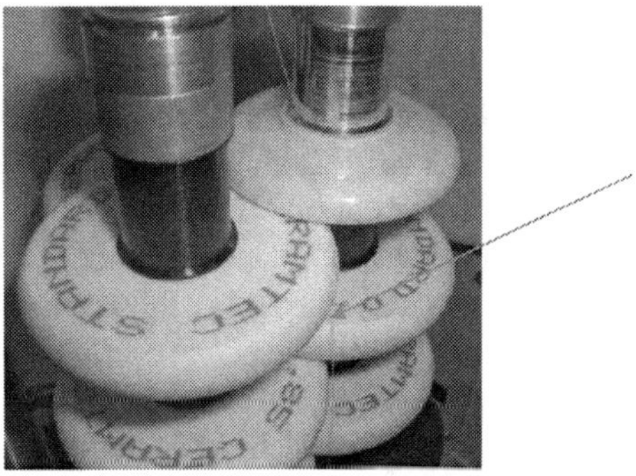

5.9 Yarn ejection from the twist unit.

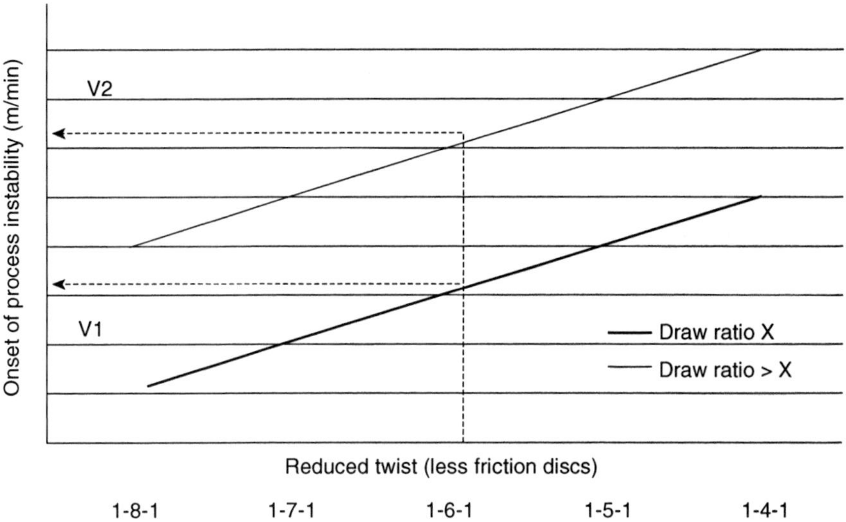

5.10 Effect of twist and tension on process instability.

increased number of friction discs or through increase in hard disc surface roughness, twist levels increase, but the surge speed reduces! Surge speed for a given yarn and machine profile is predominantly influenced by twist level and yarn tension, and it is independent of the twist application system, i.e. disc, disc type or belt at the same twist level (Fig. 5.10). On this basis, the maximum speed attainable in a draw texturing process is fundamentally limited by:

- The uppermost yarn tension that can be tolerated, above which process breaks would be too high or yarn elongation too low (yarn elongation decreases with increase in draw ratio).
- The lowest number of friction discs that can be applied (reduced twist) for attaining the required crimp value of the yarn at the higher speed.

Of course, there are other influential factors too, such as whether reduced yarn residence time on the heater at increased speed provides an adequate crimp value and whether a higher set temperature of the heater to offset loss in crimp causes higher yarn breaks or filament damage. *D/Y* settings can also influence process stability through their effect on process tension. Such parameters can be used to fine-tune a high-speed process. Twist and process tension are, however, the key parameters that influence the onset of process instability (surging) and must be optimised carefully when developing a high-speed draw texturing process.

Literature highlights the fact that twist level change through disc type and combination for given thread paths and draw ratios in a process all satisfy the same trend in terms of the onset of surging. This also applied to twist imparted by experimental belt twisting at the time.[4] Moreover, on or above speeds at which tension surging occurs, the process can also move into a random or continuous yarn slippage mode at the twist applicator;[6] this is often seen when determining the onset of process instability in the optimisation of texturing process conditions. In this phase, the high yarn tension does not recover at the transient and the yarn tends to 'bounce' at a reduced yarn angle on the friction discs. In these conditions, extremely low and uncontrolled twist is produced, which generates too low a crimp and totally inadequate twist regularity for commercial use. This slippage phenomenon tends to be more evident when using hard material, smooth-surfaced friction discs and not with softer polyurethane friction discs, where the surface friction is higher. The phenomenon may be due to insufficient torque being generated by the friction discs at the low twist/high tension phase of the tension surge for recovery to the conditions at which surging occurs. Surging has similarly been experienced with magnet spindle twist insertion[7] but in those early years of false twist texturing development, its significance could not have been fully foreseen.

So what is the likely cause of tension surging? Basic consideration points for the cause of surging are:

- Process considerations.
- Texturing zone considerations.

Process considerations affecting surging include:

- Twisting rate (torque, speed).
- Yarn tension (draw ratio, raw material draw force).
- Ballooning (yarn stability) and hence tension fluctuations.

It would appear that the characteristics of a tension transient/single surge (initiated by filament/yarn diameter increase) are the same as a single phase of periodic surging.

Texturing zone considerations affecting surging include:

- Texturing zone length (shorter zones lead to increased surge speed).
- A thread path with higher wrap angles, e.g. Barmag M profile, having a higher surge speed for a given draw ratio and friction disc configuration.

From the above, it is known that a sudden yarn diameter change (filament loop, POY splice, filament gel) brings about a single surge, which will recover to steady-state conditions. The logical question is, therefore, what could cause the same process disturbance in a periodic manner? The following factors affect periodic surging:

- Yarn tension and tension fluctuations.
- Twisting rate.
- Zone length.
- Machine profile (affects twisting torque in the pre-heater and heater zone, through frictional resistance on yarn wrap contact surfaces.

Taking all factors into consideration, it is likely that the onset of double-twisting is the fundamental source of the onset of periodic surging. This is most probably linked to unstable conditions that materialise when instantaneous diameter variations destabilise steady-state conditions for twist contraction and twisted yarn diameter relationships. Double twisting or torsional buckling has been reported in early work in friction texturing[7] and, under certain process conditions, is evident within the heater zone. Its onset is evident as twist levels are increased to an extent at which the thread line buckles into a double helical structure (Fig. 5.11).

Practical experiences concerning thread-line tension stability and thread path length do support this twisting–buckling relationship:

- Reduction in process tension (draw ratio or POY draw force variation) will increase the tendency to buckling at heater entry.
- Increased thread path angles between heater and cool zone will reduce twisting torque on the heater, reducing tendency to yarn buckling.
- Shorter heat–cool zones can be expected to reduce the amplitude in axial displacement of the twisted yarn due to tension fluctuations, reducing tendency to the onset of twist buckling.
- The frequency of the surge cycle is governed by the texturing zone length.

If indeed this is the source of cyclic surging, it is not clear why the onset of torsional buckling is twisting rate (i.e. process speed) dependent. The

5.11 Twist buckling.

answer, most probably, lies in a complex reaction between rates of filament bending, migration, tension and torsion, and should be a subject for further research investigation.

For a yarn processor, it is sufficient to know the fundamental parameters that cause process instability and limit the maximum process speed. It is normal in process optimisation to at least set the process specification speed to around 100 m/min below that at which tension surging occurs to ensure absolutly stable process conditions across the draw texturing machine.

5.3 Creel design and management

Creel management is an essential part of the draw texturing process and its design and ease of operation regarding the handling of the raw material yarn must not be overlooked in order to realise a good process performance:

- Angles of yarn wrap, yarn transport lengths and surfaces between POY package and input feed to the texturing zone can influence friction, leading to filament distortions that, in turn, cause short-term tension transients in the process. The raw yarn is almost invariably transported through aluminium tubes over distance to avoid any interference from stationary yarn ends with other running thread lines in the creel. The tubes must have a smooth internal finish to avoid filament snagging.
- High yarn friction can lead to high yarn tension during twist unit threading, leading to yarn breaks, twist unit spindle wraps and operator inefficiencies. This is particularly true with fine yarn counts. Highly polished ceramic guide surfaces are ideal for low friction in the texturing zone, but create high friction on flat yarns. In creel applications, rougher guide surfaces are necessary.
- Balloon lengths can influence tendency to filament separation on the cylindrical wall of the POY package during off-wind, leading to short-term tension faults and yarn breaks. Air interlacing of the POY in the spinning process helps to eliminate this.
- Access for creel loading must be simple to avoid any handling damage to the individual filaments of the POY yarn.
- There should be facilities to accommodate reserve POY packages. This enables air-splicing techniques to be used to join the yarn tail of the 'run-out' POY package with the reserve POY for process continuity.
- Protection layer plates are necessary below and above POY packages to avoid yarn entanglements and resultant interactive process breaks.
- Ideally, creel peg mountings should be adjustable for precise package alignment and settings for POY tube diameter.

5.12 Protection plates should not interfere with POY off-wind at full diameters.

Housekeeping disciplines are also important, particularly for high-speed operation:

- Package alignment with the yarn off-take ceramic guide is essential to maintain low yarn tension peaking during off-wind.
- Protection layer plates above and below POY layers should be clean to facilitate tail transfer and should not interfere with the off-wind of full POY packages (Fig. 5.12).
- Guides and creel tubes should be periodically cleaned to remove contaminant build-up (such as polyester oligomer, nylon 66 monomer and spin finish) in order to maintain low friction. In particular, plastic knee joints used for linking angled creel tubes should be periodically checked for assurance that the yarn contacts only the ceramic inserts at the creel tube ends (Fig. 5.13).
- POY handling should be minimised to avoid drawn- or broken-filaments that could lead to tension transients and yarn breaks in the texturing process.
- POY tube plastic end-caps should be used to avoid yarn snagging on damaged tube ends during off-wind of the inner layers of the package. Filament snagging can lead to tension transients and yarn breaks in the texturing zone. The plastic end-caps should also be inspected periodically for any damage and replaced accordingly.
- Airflow draughts (through plant door locations or positioning of air-conditioning ducts) should be avoided. These can dislodge air-spliced tails and cause interference from any loose yarn ends.

The most common types of creels (with package reserve capability for tail transfer) for raw material yarn are:

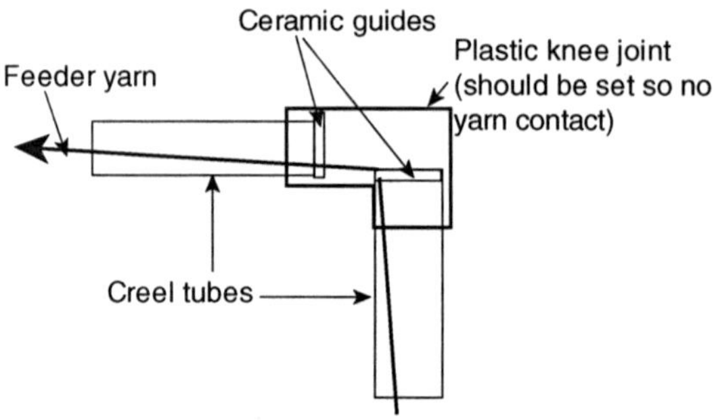

5.13 Plastic knee joints for coupling creel tubes.

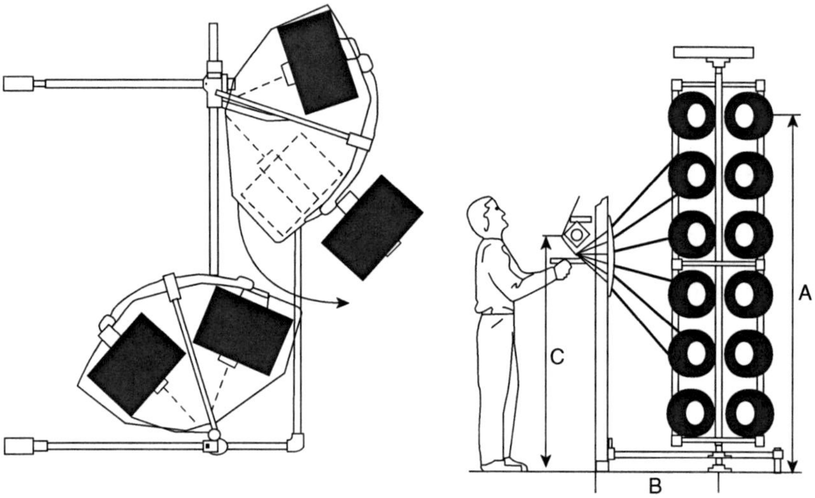

5.14 Six-high swing-out creel example (Rieter-Scragg).

- Vertical, non-rotary, swing out POY support pegs for loading, usually six package-high (Fig. 5.14).
- Rotary creel, three/four package-high, with low height to reduce POY lifting (Health and Safety legislation) and access to all POY support pegs through creel rotation (Fig. 5.15).
- Four-high rotary creel, which is commonly used in the industry (Fig. 5.16).
- Trolley creels, which slide out from the machine for ease of access.

For process continuity, yarn ends between the running POY and reserve package can be joined using the tail facility on the reserve package. Originally, this used to be done by carefully hand-tied knots but today, purposely designed air splicers are used. These entangle the filaments to create a low diameter, high-strength knot and, in some designs, cut the loose yarn ends at the splice. A low diameter knot and minimum filament disturbance

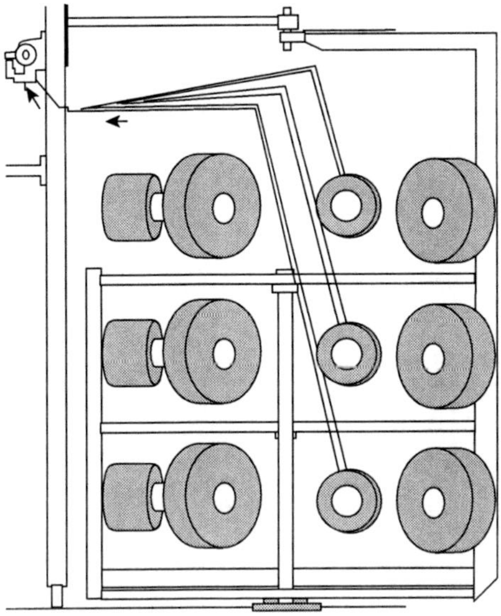

5.15 Three-high rotary creel for ease of POY handling (Rieter-Scragg).

5.16 Four-high rotary creel (Barmag) with protector layer plates.

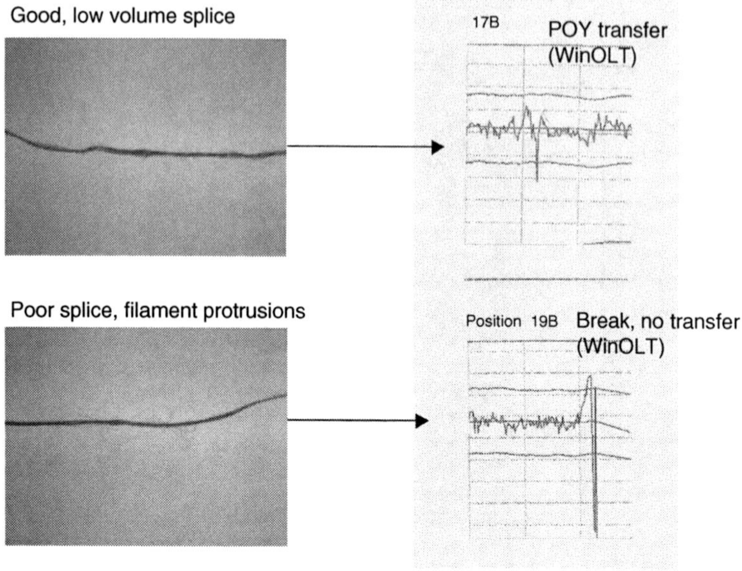

5.17 Effect of splice quality on process tension transient and associated break.

before and after the knot to maintain the parallel nature of the filaments is important; large disturbance to the filaments can lead to process tension transients in the texturing zone and subsequent yarn breaks. This is demonstrated in Fig. 5.17 through the monitoring of process tension during splice transit using the TEMCO WinOLT online tension system.

Applying tail transfer demands much discipline:

- The splice must be firmly held away from the running package to avoid any interference, but must be released at package changeover without any significant tension disturbance. This is normally done by the retention and simple release of the splice using a twin nylon monofilament clamp or a soft Velcro patch located on the rear of the protector plates in the creel.
- The splice should not have protruding filament ends; these are automatically cut by the air-splicer or carefully cut manually.
- Filaments about the splice should remain parallel; filament loops create large tension transients and often process breaks.
- The splice diameter should be as small as possible; often it helps to roll the splice gently between the fingers after air splicing to assist its transit through the texturing zone.
- Outer layers of the reserve package should not be disturbed or damaged through bad handling practice.

Automatic monitoring systems are available for creel management performance in terms of tail transfer efficiency and package change links with process tensions.[8]

5.4 Package build specifications

High-speed off-wind requirements of up to 1500 m/min in downstream processing impose heavy demands on textured yarn packages. Package build specifications require careful optimisation to achieve trouble free off-wind of textured yarn packages, which have been built historically on the 'random wind' principle to avoid escalation of machine costs. The yarn and its crimp character can also strongly influence the off-wind performance of textured yarn packages.

The general basic rules to obtain good off-wind performance on textured yarn packages for direct use in knitting, warping or weaving is to maximise the packing density. Typical package densities are around:

- PES 167dtex f32–620 g/L.
- PA66 17dtex f5–720 g/L.

High package densities minimise tendencies to yarn off-wind tension peaking and yarn snagging on the package, which would otherwise have an adverse effect on efficiency and product quality in downstream processing. Increased wind tension increases the package density, but there is an uppermost level that can be used due to the onset of:

- Tube crushing due to the contraction force of the yarn, particularly with nylon 66 (Tube specification plays an important role here, for example, board quality and wall thickness).
- Webbing (yarn overthrows on the package taper wall).
- Package aesthetics (shape).
- Broken filaments (particularly concerning microfilaments) due to tension peaking at traverse reversal.
- Dye fleck in the fabric due to high tension peaking and resultant drawing of the yarn at traverse reversal.

To satisfy packing density requirements, reduced traverse rate and hence wind angle on the package ensures high wind packing density with lower tendency to tension peaking. In practice, it is common to operate at wind angles between 12 and 14 degrees, i.e. 24 to 28 degrees crossing angle (Fig. 5.18).

Packages wound by the random wind principle are subject to an effect known as 'ribboning'. A ribbon pattern, due to yarn being laid over or immediately adjacent to the previous wind, is formed whenever a package

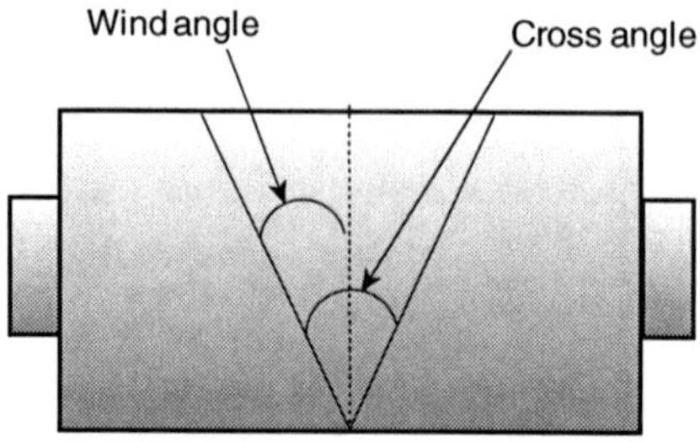

5.18 Wind- and cross-angles in package build.

revolves through an integer number of revolutions during one complete traverse cycle:

$$D = \frac{S \times 1000}{t \times \pi \times n} \quad [5.2]$$

where D = package ribbon diameter (mm)
S = package surface speed (m/min)
t = traverse rate (cycles/min)
n = 1, 2, 3 etc.

Machine manufacturers build in a disturbance systems to vary the traverse rate during package build to minimise the effect of ribboning, which otherwise tends to cause yarn snagging during off-wind. Such anti-patterning can usually be set in terms of cycle magnitude and time by the machine user.

Adjustable package edge disturbance systems via microprocessor control are necessary to avoid high package edge densities caused by concentrated yarn build at traverse turn-round. There are two basic variations that can be applied to the disturbance:

- Axial dwell (percentage dwell time at full traverse stroke).
- Axial displacement (shortening of traverse stroke to within pre-set levels on both ends of the package build).

From these, the user has the opportunity to select both edge disturbance patterns and amplitudes (Fig. 5.19). Of these two variables, axial dwell has been found to have the most effect on package off-wind.[9,10] Low dwells for softer package edges help off-wind, but there is a threshold beyond which webbing or yarn overthrows can occur.

Perhaps the leading package build support in draw texturing over the last 15 years has been the Rieter-Scragg APS winder (Advanced Package Support).[9] The APS concept is depicted in Fig. 5.20. The APS cradle has the following features:

- A rigid construction to minimise vibration (cradle vibration can cause yarn overthrows at traverse reversal).

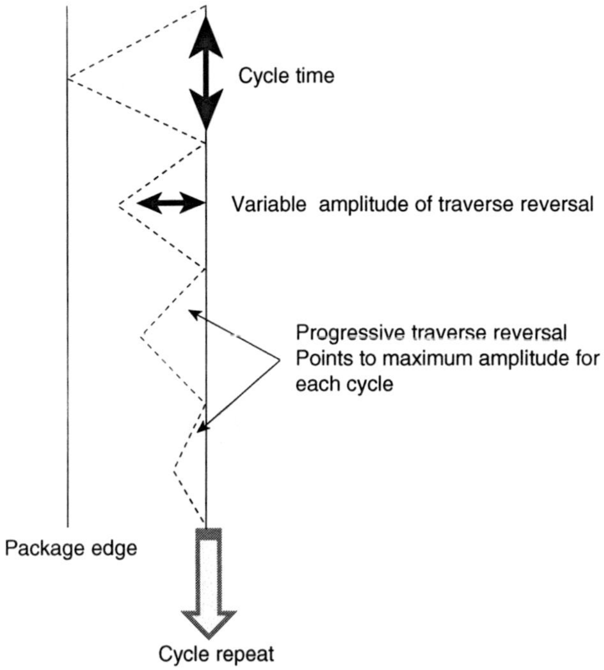

5.19 Concept of edge disturbance for programmed change in traverse reversal.

- A unique package holder rotating system to maintain uniformity in package drive force on the cylinder through package weight compensation. No springs are required to maintain constant package force on the drive cylinder.
- During traversing, the yarn has contact on the drive cylinder for dampening purposes at traverse reversal. This allows low wind angles to be applied for high package build density and off-wind performance (Fig. 5.21).

It is often overlooked that the yarn and crimp character can influence package off-wind performance:

- Loops in the textured yarn can cause filament snagging and tension peaking on the surface layers of the package.
- Microfilaments are particularly problematic in terms of filament snagging and there are limits regarding the uppermost package density that can be applied. For microfilaments, the high wind tensions necessary for high package density can cause broken filaments or short lengths of drawn yarn at traverse reversal. As a result, package densities typically around 600 g/L apply, depending on yarn type.

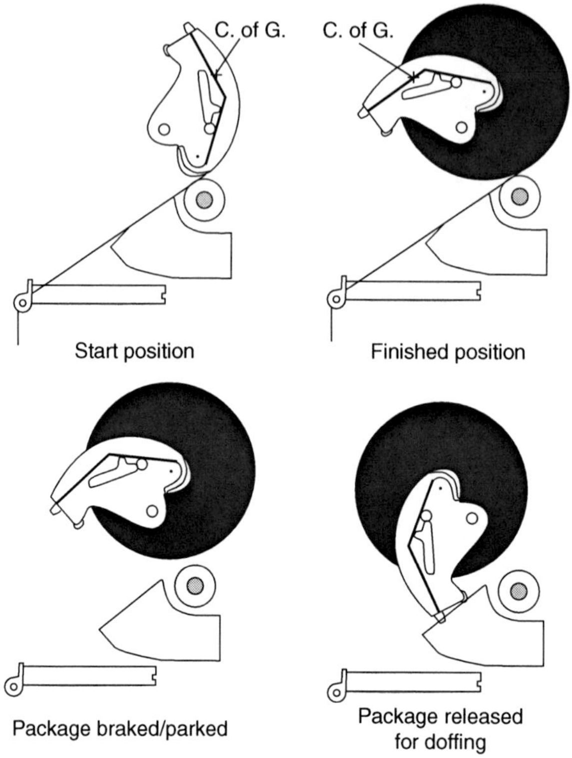

5.20 Working principle of the Rieter-Scragg APS winder.[9]

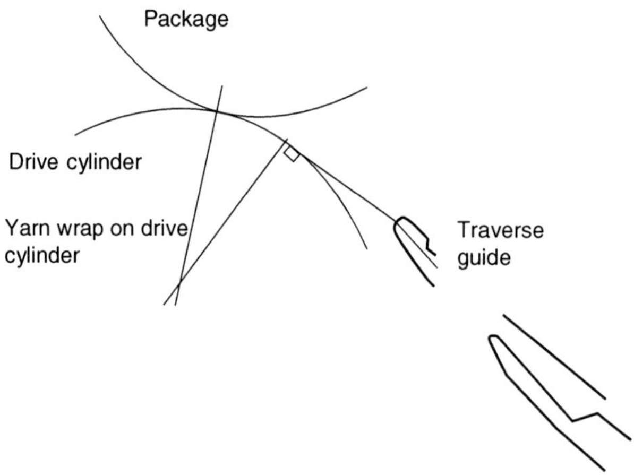

5.21 Yarn contact angle on package drive cylinder on APS winder.

- Air intermingling facilitates trouble free package off-wind, but its application is dependent on yarn end-use. For microfilament yarns, package off-wind performance can be improved through the use of low pressure air intermingling in the pre-texturing zone thread path.[11] This predetermines filament migration during twisting, reduces loop formation and gives a more compact crimp appearance in the yarn. As a result, any tendency to filament loop snagging during package off-wind is much reduced.

Where high package densities are necessary to support high-speed off-wind of draw textured yarn packages in direct downstream applications, package build requirements are different for packages used for colour application via subsequent package dyeing processes. For dye-packages, higher wind angles and much lower package densities are necessary for uniform dye penetration throughout the wind layers in the dye bath. Package densities need to be particularly low for the finer filament yarns, where the filament surface areas are higher and uniform dye liquor penetration more difficult.

There are two types of plastic perforated dye tubes that are usually used for polyester textured yarn dye packages, viz. compressible and rigid (example, Fig. 5.22). The tubes are designed to be inter-lockable on the dye stem so that dye penetration is achieved by a programmed flow of dye liquor through the yarn dye packages, avoiding any package bi-passing of the liquor. Compressible tubes are compressed approx. 20–25% on the dye stems before dyeing.

For polyester yarn, typical package build conditions for direct dyeing are

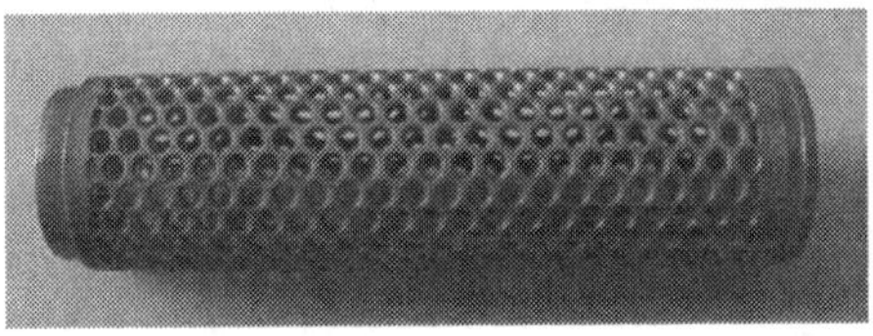

5.22 Non-compressible plastic dye tube example.

Table 5.1 Package build conditions for direct dyeing

	Tube type	
	Compressible	Non-compressible
Density (g/L)	340	420
Wall taper (degrees)	Zero	18
Wind angle (degrees)	32–38	34
Approx. weight (kg)	2.2	3.0

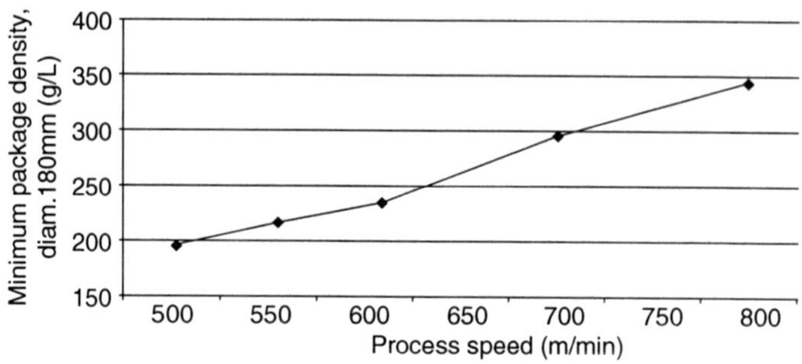

5.23 Example of process speed limitations for nylon 66 yarn for achieving low package densities.

shown in Table 5.1. During dyeing, the yarn shrinks on the package and it is important to select wide amplitudes for traverse edge disturbance when optimising the package build. This provides soft package edges within which the yarn has an opportunity to shrink and dye uniformly with the rest of the build, otherwise stripes corresponding to the traverse reversal points will be evident in the textile fabric. For optimum builds, polyester package-dyed yarn can usually be used directly in the downstream process. Nylon 66 yarns, however, are more difficult for direct use; they have higher yarn contraction force and shrinkage in dyeing, and to retain crimp values, package densities need to be extremely low. As a result, nylon 66 dye packages are commonly wound on metallic dye spring tubes at extremely low package densities of around 200 g/L. To achieve such low densities, wind tensions are virtually zero and process speeds are very much restricted (Fig. 5.23). Dye packages wound on metallic dye springs have to be rewound onto paper tubes in a subsequent winding process after dyeing for downstream use.

More recently, new machine developments, which include package build considerations, are evident from the remaining European manufacturers. However, because of their high manufacturing costs, these tend to be aimed at specialised properties and markets, such as process flexibility needed for a quick response to customer demands and dye package manufacturing. For countries with high volume outputs, for example China and India, package build technology is still based on the above know-how, developed in Europe during the 1970s to 1990s. New package build systems are discussed further in Chapter 6.

5.5 Oil application systems

Lubricant (coning oil) is applied to textured yarn prior to package build in order to:

- Reduce yarn contact friction in downstream processing.
- Improve package off-wind through reduced inter-yarn friction.
- Eliminate electrostatic charges generated through downstream frictional contact.

Typically, 1.5–2.5% coning oil is applied to textured yarn. It is not necessary to apply coning oil during the processing of dye packages as it would be washed out in the dyeing process. For package dyeing, lubricant is added as an after-process in the dye vessel, or during rewinding of the packages.

Methods of lubricant application in the draw texturing process are through:

- Contact roller between the output feed and take-up winder (oil pick-up adjusted by roller speed). Such lick rollers can be smooth or serrated (Barmag).
- Contact roller on each take-up deck prior to the traverse mechanism.
- Oil jet, gravity feed system (Metoil, Rieter-Scragg).

Oil application systems and their thread path geometries are important aspects of the machine design, as frictional yarn drag on oil rollers can be problematic, especially for high viscosity oils with low splash properties if the arc of yarn contact is high. Commonly encountered problems are:

- Yarn breaks; wrapping on the output feed rollers due to pre-oiler low yarn tension within the doffing sequence on automatic doffing systems. It is common for the yarn to be automatically lifted away from oil roller contact during the doffing sequence.
- Package density operating window limits; low pre-oiler yarn tensions occur due to frictional drag on finer filament yarns, causing yarn flutter and yarn breaks through output feed wraps.
- Torque yarns with high residual torque cannot be produced where there is high frictional drag due to the tendency to self-twist in the pre-oiler thread path at low yarn tension.
- Frictional drag can cause sufficient filament separation to create package off-wind snagging on fine filament count yarns.

The above phenomena are illustrated, together with means of their elimination in Fig. 5.24.

Other means of reducing the low tension problem prior to the oiler are:

- To fit oil rollers per take-up deck after the guide or bar before the traverse. Here, the filaments are under higher tension and less problematic regarding drag on the roller. Such designs are commonly used on draw texturing machines, but have the disadvantage of added component cost and tend to create more oil splash as the yarn traverses over the oil roller.

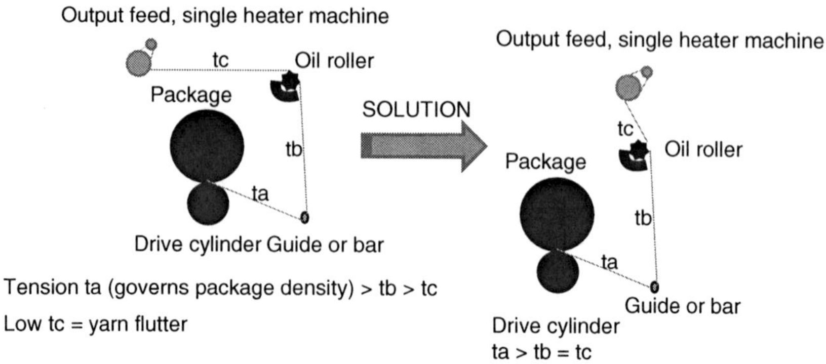

5.24 Design issues that limit the operating window for the take-up zone yarn tension and package density.

- To use oil jets, as introduced by Rieter-Scragg in their Metoil system, where the oil is metered by a simple gravity feed system. This system proved highly successful in both polyester and nylon applications and was particularly advantageous when processing torque yarns.

5.6 Automatic doffing systems

Over the last 15 years, the industry has experienced growth in the sale of texturing machines with automatic packing doffing facilities in high labour cost countries. The growth has been predominantly in polyester texturing, where yarn linear densities tend to be relatively high and the period between doffs short, i.e. for processes that demand higher machine operator input.

The advantages of automatic doffing systems are, however, not only associated with machine operator cost reductions. They add:

- Equi-length packages for warp and dye-package circular knitting for automotive upholstery end-uses.
- Package weight consistency for dye-packages (better dye shade regularity, within and between packages).
- The additional feature for applying a consistent 'inner-waste' wind on dye-packages to avoid dye-shade difference on the inner wind layers of the packages, i.e. a double tailing capability over a pre-set wind time for inner package layers.
- A platform for further robotic handling links for packages, i.e. transfer to transportation robotics for direct packing, etc.

The disadvantages of automatic doffing systems are, however:

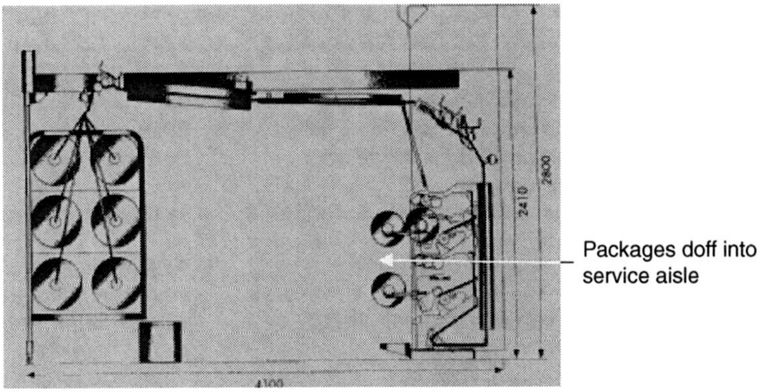

5.25 Example of in-board doffing, single machine side of ICBT FTF15 E2 HT machine.

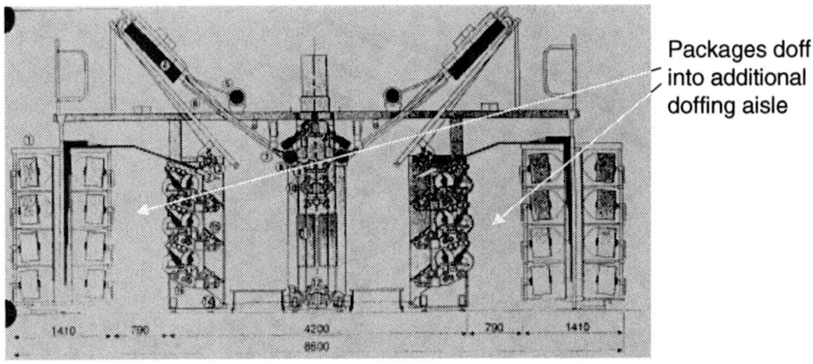

5.26 Example of out-board doffing, Murata 33J Mach crimper.

- Additional machine cost (lower cost through alternative manual doffing with simple doff timers per position can be adequate for longer doff times).
- Usually additional floor space is required.
- Additional machine maintenance complexity and cost.

Package automatic doffing systems tend to be either designed within the machine main frame (in-board doffing, Fig. 5.25) or remotely from the machine (out-board doffing, Fig. 5.26).

In-board and out-board doffing systems have the comparative disadvantages outlined in Table 5.2.

In general, for automatic doffing, it is possible to select between random or non-random doffing modes across the machine, the choice of which

Table 5.2 Comparative disadvantages of in-board and out-board doffing systems

In-board	Out-board
• Three-deck take-up requires operator trolley for machine threading (2-deck more floor space per machine output)	• Creel cannot be accessed from service aisle
• Yarn storage is in the service aisle, inhibiting operator access	• Increased floor space
• Shorter thread path in the take-up zone (potentially higher tension peaking at traverse reversal for yarn with low extensibility)	• Potentially higher noise emission through encapsulated service aisle
• Maintenance access	

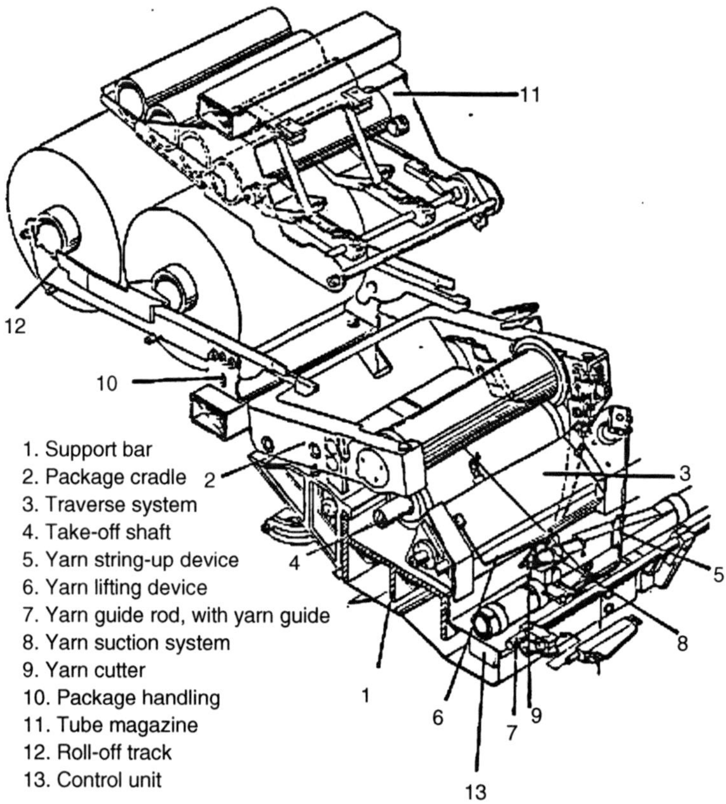

5.27 Example of automatic doffing module (two-pack storage) (Barmag AFK) (Illustration © Oerlikon-Barmag).

is dependent on the preferred economics of the process. In the random mode:

- The doff timer is activated for individual package set runtimes and packages are doffed at random.
- Where process break rates are relatively high, it is the preferred option, i.e. output efficiency is maintained at high level through rethreading broken yarn ends with full utilisation of machine positions.

In the non-random mode:

- Positions cannot be rethreaded after yarn breaks until immediately before machine doffing is due, to avoid cost of low weight, downgrade packages; alternatively, the processor can re-thread at break producing low weight packages. Consideration for use is, therefore, a balance between output efficiency and downgrade package costs.
- Consideration has to be given to the maximum number of positions that can be simultaneously doffed on the machine; otherwise, the high volume of waste yarn in the suction can block the suction tubes.
- For extremely low process break rates it is an easier, systematic and manageable system, which can be more suited to synchronisation with robotic handling devices for package removal.

Automatic doffing systems comprise cassettes for reserve tubes and textured yarn package storage (Fig. 5.27). The time required for exchanging textured packages at doff is typically 20–30 seconds. It is important in their designs that large yarn wrap angles are minimised during the transfer of the yarn from suction to capture at the end-cap of the tube, otherwise upstream low tensions can result in yarn wrapping at the output feed prior to the take-up unit. This is also particularly important in the processing of low-density dye packages, where wind tensions are low.

5.7 References

1 J. E. Obetz, Texturising Lubricants and Finishes, *Textile Asia*, 20–24, Nov. (1973).
2 M. J. Schick, Friction and Lubrication of Synthetic Fibres, Part I: Effect of Guide Surface Roughness and Speed on Fibre Friction, *Textile Research Journal*, 103–109, Feb. (1973).
3 C. Atkinson, J. Parnaby *et al.*, MMF Processing: The Balance Between High Speeds and Yarn Quality in Draw Texturing, *International Textile Machinery*, 35–38 (1981).
4 G. E. Isaacs, C. Atkinson, *Optimisation of Process Performance and Yarn Quality at Current Texturing Speeds*, Developments in Texturing, Shirley Institute Publications S 26 (1980).
5 C. Atkinson, Machine Configurations in High-Speed Drawtexturing, Rieter-Scragg Ltd, *Chinatex, Shanghai*, Jun. 10–16 (1984).

6 J. W. S. Hearle, L. Hollick, D. K. Wilson, *Yarn Texturing Technology*, Woodhead Publishing Ltd. (2001).

7 S. Backer, D. Brookstein, Material Process Interaction During False Twist Texturing, *Journal of Applied Polymer Science: Applied Polymer Symposium*, 31, 63–82 (1977).

8 U. Wagner, M. Herzburg, High-Speed Texturing with highest Flexibility, *Chemical Fibers International*, 54, 332, Oct. (2004).

9 C. Atkinson, High-Speed Draw-Textured Package Build and Offwind, *Chemiefasern/Textilindustrie*, 167–173, Mar. (1985).

10 K. Fischer, Aufbau und Kontrolle von Kreuzspulen, *Chemiefasern/Textilindustrie*, 240–247, Apr. (1985).

11 C. Atkinson, S. Müller-Probandt, Opportunities for Further Optimisation of the Primary Zone in False Twist Texturing, *Man-made Fiber Year Book 2006*, Chemical Fibers International.

6 Newer yarn texturing machine designs

Abstract: This chapter reviews more recent yarn texturing machine designs. It discusses factors affecting machine design and example of new designs.

Key words: false twist texturing, machine design.

6.1 Factors affecting machine design

The relocation of draw textured yarn manufacturing to low labour cost countries in predominantly the Far East and Asia has seen the support for machine manufacturing and machine components subsequently rapidly grow in China and to a lesser degree in India. Where labour costs are low, automatic doffing of draw textured yarn packages on cost saving grounds is not a key consideration and machines produced in China and India tend to be based on, and copied from, machine designs commonly manufactured in Europe in the 1980s. There are, however, signs that automatic doffing is starting to emerge in polyester draw texturing in India.[1]

Moreover, process speeds are restricted and cannot be markedly increased due to the onset of instability, which is a fundamental speed limitation of the draw texturing process. Also, finance costs associated with machine investment are major consideration points in draw textured yarn manufacturing. As a result, basic, low-price draw texturing machines and components manufactured in China are adequate in supporting local, large volume textured yarn manufacturing.

In-depth machine design and process know-how for draw texturing arguably still, however, remains for the moment within the countries that were previously prevalent in the industry. In these countries, there remains a market segment requiring a quick response to the supply chain with higher value-added, purposely-designed, often small quantity yarn products where the variables are ranging from yarn count, filament fineness and yarn type to colour. In the case of nylon, draw textured yarns commonly are

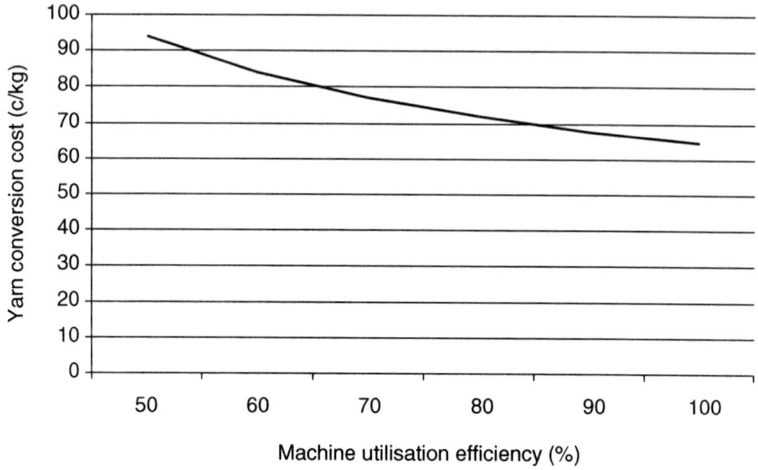

6.1 Influence of machine utilisation on conversion costs (Eurocent) for nylon 66 44 dtex f34.

subsequently combined with elastanes for high elastic recovery in knitted garments, particularly in ladies' hosiery. Moreover, 'supply on demand' response for cost effective, minimum yarn stock-holding operations must be supported by highly flexible process machinery, capable of supplying exact yarn quantity requirements at the highest process speeds, machine-utilisation and -conversion efficiencies, and indeed optimum yarn qualities.

Traditional line shaft machines in draw texturing are, however, somewhat restricted in satisfying these demands:

- Where limited volume, high value-added products are required, full machine utilisation is not always possible, which together with frequent process specification change needs, severely restrict output efficiencies and adversely affect process costs (Fig. 6.1).
- For sensitive yarns, frequent process specification changes to other yarn counts and raw material sources often necessitate heater- and friction disc-cleaning between individual processes with resultant lengthy machine side- or full-machine stoppages, negatively influencing machine utilisation efficiencies and process costs.
- Random wind technology applied to package build on line shaft machines often needs supporting with air interlacing technology for trouble-free package off-wind in down-stream processes, especially concerning fine filament yarns. As a result, shorter tube- and traverse-lengths are sometimes used to improve off-wind performance, but in these cases maximum package weights are limited.

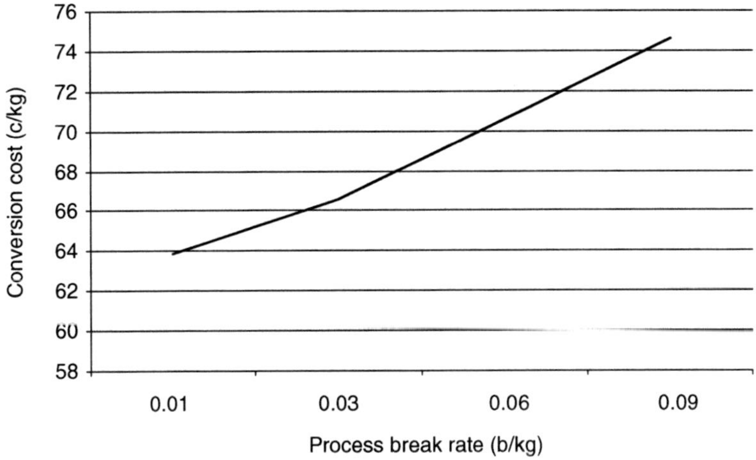

6.2 Influence of end breaks on yarn conversion cost (Eurocent) for nylon 66 44 dtex f34.

- Where yarn producers have no sample machines that correspond to the exact thread path of their production machines, they are unable to rapidly transfer new products from development to efficient manufacturing without high costs incurred due to the excessive production machine downtime necessary for the development phases that involve the utilisation of only a few processing positions.
- The majority of modern line shaft machines are either equipped with standard vapour phase or high temperature (HT) heaters; vapour phase heaters do not satisfy the range of set temperatures required for say the smaller throwsters to texturise nylon, polypropylene and more recently PLA yarns (see chapter 13) without a change of vapour phase medium. On the other hand, HT heaters are known to be problematic concerning process efficiencies on the finer count nylon yarns.

Conditions of older line shaft machines and sometimes their profiles can dramatically increase process costs due to poor product quality conversion and high break rates, particularly on sensitive finer count nylon yarns (Fig. 6.2). As a result, process speeds are often restricted.

Together with the above process cost factors, there are other machine design areas and opportunities where further significant process cost savings can be made in draw-texturing:

- Power consumption savings through heater design and drive systems that only consume energy on running positions.
- Reduced spare parts requirements.
- Lower floor space utilisation through machine design.

Table 6.1 Features of new machine designs introduced by remaining European manufacturers

Process flexibility	Design concept for a broader range of products Smaller quantities for niche markets Multi-process specifications across machine for simultaneous production of different products
Process costs	Largely automatic doffing and equi-length packages Energy conservation Component costs
Ergonomics	Floor operational Automatic yarn feeds
Environmental issues	Energy conservation Low noise emission

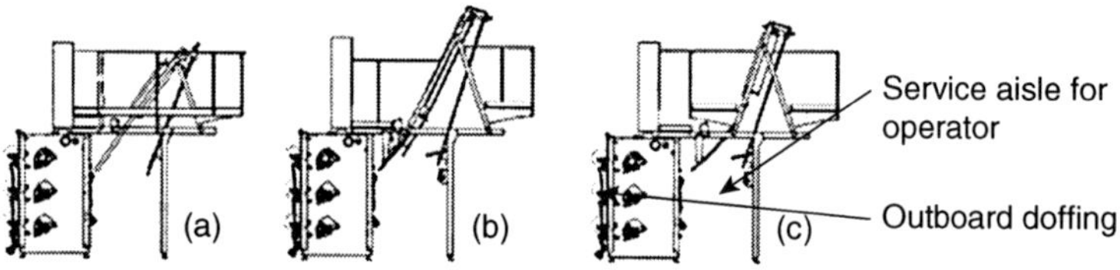

6.3 MPS machine cross section with options. (a) MPS-V-MAX with HT heater and cool tube (b) MPS-V with cool track and vapour phase heater (c) MPS-V-HTI with cool track and High temperature HT heater.

- Integration of the air covering process into draw texturing for elastane combination yarns.

With this awareness, remaining machine manufacturers in Europe have focused on new machine designs for largely local demand,[2] concentrating on the features detailed in Table 6.1.

6.2 Examples of new designs

The Oerlikon-Barmag MPS-V-MAX machine (Figures 6.3 and 6.4) with a maximum mechanical speed of 1500 m/min, is a good example of the recent focus on process cost saving and flexibility.[3] Many well-proven features from the Barmag AFK draw texturing machine with outboard automatic doffer have been incorporated on this machine. The machine was launched in 2003 with numerous novel features:

- Godet/separator roller feed systems replace standard nip roller or apron feeds for low maintenance costs, eliminating the need for periodic nip

6.4 Double density MPS machine.

roller grinding due to wear or apron replacement. Yarn slippage at the feed system due to nip roller wear is also claimed to be avoided.

- Automatic yarn feed for threading through the second heater to the automatic doffer using pneumatic suction devices saves labour.
- The machine can be equipped with electrical resistance 1.0 m length HT heaters or standard 2.0 m vapour phase heaters. In the former case, the yarn guide sleeves can be interchanged with full yarn contact sleeves for wider process operating windows to accommodate alternative yarn types. Energy consumption from this contact heater option is claimed to be reduced by 57% over that of the vapour phase heater.[3]
- An ATT take-up system (Advanced Take-up Technology), which comprises stepping motor and gear belt drive, and sophisticated software controls[4] provide a means of maintaining constant packing density of the yarn during reduced traverse stroke for a range of package wall taper angles. Moreover, novel package shapes at low package density can be produced to improve dye liquor penetration uniformity and dye shade consistency in package dyeing (Fig. 6.5).
- Modular process specification control enables small quantities of different products to be produced simultaneously.

Standard build With ATT system

6.5 Dye package with novel build on ATT system.[4]

- Floor operations are provided for operators when servicing the texturing zone.
- Individual drive motors for the twist units improve speed consistency and reduce noise emissions.
- Fitting adjustable cool tubes for increased process speed is an option.

Godet feed systems have also been introduced to other machines in the Oerlikon Barmag product range aimed at energy and maintenance cost savings:

- Their FK6-1000 based on texturing zone designs from the 1980s can now be supplied as an eFK machine with godet feeds (with the exception of the output feed to the take-up, where the conventional line shaft drive has been retained). A multi-spindle version for multi-ply of fancy yarn production is also available with claimed energy cost savings of around 40%.
- Their previously introduced COCOON machine was equipped with godet feed systems, and with the ATT winder that has claimed energy cost savings of around 25% against the FK6 machine. The machine is based on manual doffing. Maintenance costs are also improved, enhanced by longer periods between maintenance needs and, as a result, machine downtime.

A further interesting machine that entered the market in recent years is the DP3-FT, a single heater draw texturing machine manufactured by SSM, Switzerland (Fig. 6.6).[5] The machine is predominantly aimed at the nylon and polypropylene markets, and takes process flexibility to the ultimate with each position being individually controlled. The machine also features an integrated texturing and air covering process that offers high potential for reducing the processing costs of air-covered elastanes against costs

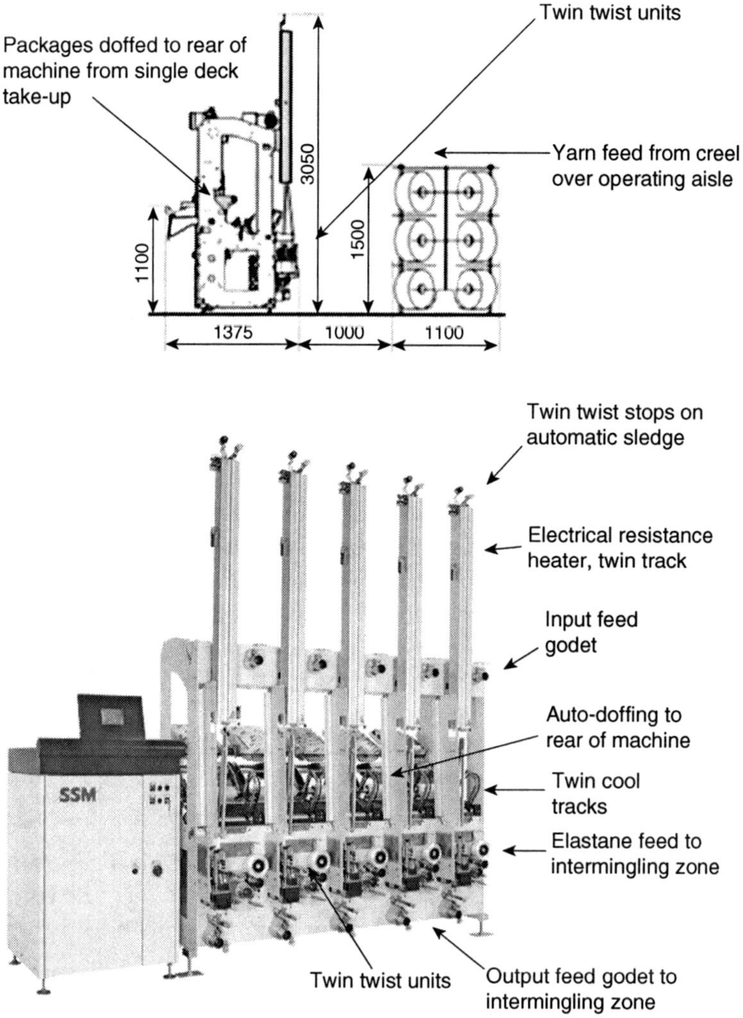

6.6 SSM DP3-FT combined draw texturing and air covering machine.[5] (Photo supplied by SSM, Switzerland.)

associated with the standard two-process route, due to savings in capital investments, energy, consumables, labour and floor space.

The DP3-FT is designed around universal feed controllers for each processing position, enabling the optimum in process flexibility in draw texturing. It also incorporates the well-proven SSM Digicone precision wind system from SSM, for optimum package-build performance in all end-uses. Other key features are:

- Purposely designed electrical resistance heaters ensure temperature profile consistency between heaters and along the yarn-to-track contact lengths; they can accommodate fine to medium count nylon and polypropylene yarns.

- The machine has a mechanical speed capability of up to 1200 m/min
- The facilities to apply alternative draw ratios and speed during start-up for trouble free threading of sensitive yarns are provided. This feature can be particularly advantageous for fine count nylon and polypropylene yarns.
- Selection of a lower speed when the yarn is running in the aspirator during the doffing cycle enables totally trouble-free doffing.
- A yarn thermal protection system is featured in the texturing zone at positional stoppage, which operates when online selected quality parameter thresholds have been exceeded within a textured package, manual stoppage of the position, or an elastane break in the combined process occur. As a result, there is no need for operators to re-thread the texturing zone after positional stoppage.
- A two-fold process facility using twin-track heaters for combining two textured yarns for fancy effect or enhanced yarn softness and handle in fabric is provided.
- Through its compact design, all key components, such as the creel feed exits, feed rollers, twist units, air jets, doffing aspirators and positional control buttons are readily accessible from an operator standing position at each processing position.
- TEMCO open/close twist units for easy threading and TEMCO WinOLT on-line tension monitoring are provided.
- Godet/separator roller feed systems give low-cost maintenance.

The concept of integrating the covering process for elastane into the draw texturing process, predominantly for nylon yarns, has, in recent years, become more common. Many machine manufacturers now incorporate an optional facility to combine elastane via air intermingling. Giudici, Italy, who has many years of experience in draw texturing nylon yarns and being located in the heart of the ladies' hosiery manufacturing industry, is no exception here. On their combination machine, they have introduced a novel means of automatic elastane package transfer using air splicing to reduce positional downtime due to elastane package run out (Fig. 6.7). The machine can also be equipped with a short secondary heater for wider scope in producing desirable yarn crimp effects.

Despite novel machine developments, volume sales of the latest automatic flexible machines can currently be considered to be limited:

- Speed restrictions due to the fundamental problem of process instability cannot be eradicated. Thread path modifications in the texturing zone bring about small speed increases but these can be considered insufficient to offset marked increase in machine investment costs.

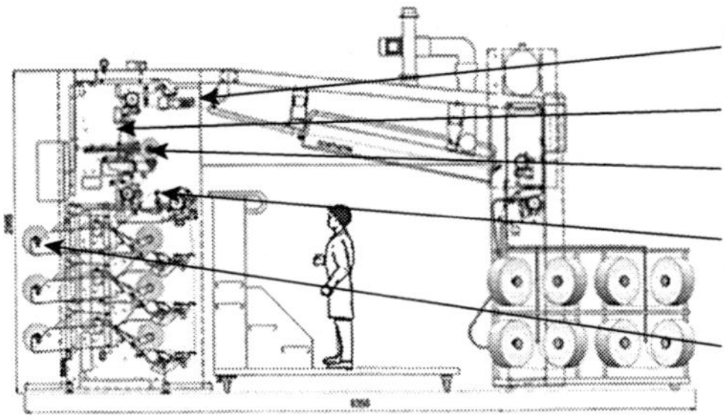

6.7 Giudici TG30 AE 'Combi' machine. (Photo from Giudici, Italy.)

- Low labour costs in the main market, China, do not yet support the need for automatic doffing.
- Standard draw texturing machinery can be manufactured at significantly lower costs in China and the higher investment costs for fully automated machinery cannot be fully justified.
- Large volume yarn manufacturing prevalent in China and India does not support the need for enhanced process flexibility.
- Conventional covering remains the main method of combining elastane in the Far East at this time.

It can therefore be assumed that perhaps, with the exception of the Oerlikon-Barmag eFK and COCOON machines, there is currently limited scope for large volume sales of the latest machine developments in the market. However, these machine developments will undoubtedly set a platform for the future, as market demands adjust globally to ever increasing wider demands.

6.3 References

1 www.himson.com/ath-12-f-v.html.

2 U. Wagner, M. Herzburg, High-Speed Texturing with Highest Flexibility, *Chemical Fibers International*, 54, 332, Oct. (2004).

3 Oerlikon-Barmag, *MPS As Creative as the Market*, Publication OBA 307 e/7/2007.

4 F. Herwegh, ATT Take-up for all Requirements of Modern Texturing, *Man-made Fiber Year Book*, Chemical Fibers International (2006).

5 *Technical Brochure, DP3-FT*, SSM Schweiter Schärer Mettler AG, Switzerland.

7 Air jet intermingling in yarn texturing

Abstract: This chapter reviews air jet mingling in yarn texturing. It discusses the intermingling process and its implications for downstream processes.

Key words: false twist texturing, air jet mingling.

7.1 The concept of air jet intermingling

Draw textured yarns comprise continuous filaments with crimp, which lack inter-filament cohesion. Filament separation from the yarn in some instances can be problematic in downstream processing due to:

- Filament snagging on adjacent wind layers during package off-wind. This can create yarn tension peaks, broken filaments and sometimes yarn breaks. Depending on the yarn property, tension peaking can cause bands in circular knitting, and weft stripes in weaving during high speed off-wind where yarn feed accumulations are not used. Broken filaments can also lead to 'slubs' in fabric.
- Filament trapping during needle penetration/yarn release in knitting.
- Filament snagging (warp-weft contact) during passage of the weft yarn through the warp shed in weaving.

To overcome these tendencies, inter-filament cohesion is imparted to the draw textured yarn by the following processes:

- A subsequent twisting process, which is a slow-speed, high-cost operation but provides a smoother better handle fabric depending on the twist insertion level applied.
- Sizing of warp yarns for weaving application, which again is a high-cost process with environmental implications (the size has to be washed away from product).

- Air intermingling, whereby a cold airstream is applied to the yarn on the draw texturing machine, imparting filament entanglements along the yarn axis.

The principle of air intermingling is also utilised for the assembly of yarn combinations in the draw texturing process; for example:

- Textured yarns with elastomers (high elasticity yarn).
- Differential yarn dye-uptakes to produce marle effects.
- Differential yarn shrinkages to produce crepe fabric effects.
- Simple combinations of yarns, typically two- to six-fold.
- Pre-texturing zone filament entanglement for low dpf and microfilament yarns to promote filament migration within the twisting mechanism.

7.2 The intermingling process

The term intermingling tends to be globally used for the air entanglement of continuous filament, draw textured yarn. There are, however, synonyms of the term 'intermingling' still used in the industry:

- Tangle lacing.
- Tangling.
- Interlacing (usually associated with 'flat' or POY yarns).
- Co-mingling (intermingling of two yarns).

So what is the mechanism of air intermingling, which has become increasingly used in the draw texturing process? This is described in Fig. 7.1, whereby:

- Textured filament yarn is subjected to a turbulent cold air-jet, which impinges on the filaments radially within a confined yarn path channel.
- The air flow opens up the filaments and the filaments are intertwined to form compact sections (nodes).
- Opened and mingled (node) sections form sequentially along the yarn axis as the yarn continuously moves through the jet, the periodicity of which is largely governed by the jet dimensions.
- The resultant intermingled yarn comprises visible nodes separated by open crimped filament lengths.

The consistency of intermingle node frequency is important in the yarn, particulary where high intermingle strength is required:

- Open, non-intermingled lengths can create appearance differences in the fabric.
- For multi-colour marle effects, through yarn combinations, inconsistencies show as stripes in the fabric.

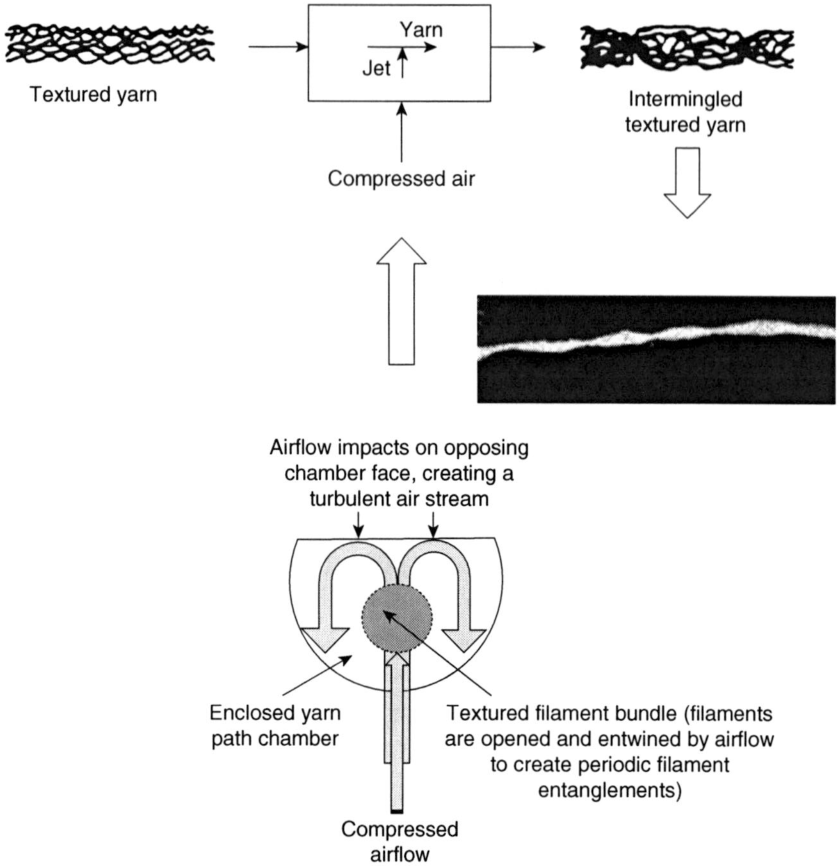

7.1 Principle of intermingling.

- When combining elastane monofilaments with textured yarn as a combined process or in a subsequent intermingling process, there is a tendency for the elastane to show as an intermittent streak in the fabric.

In many cases, low intermingle strength and good intermingle consistency are required so that the intermingle nodes are disentangled and eliminated under tension in the fabric finishing, in order that the nodes are not evident as pin-hole appearances in certain woven fabric designs. For such applications, air intermingling is applied simply to facilitate package off-wind and to eliminate any tendency for filament snagging in the downstream process. In general, most draw textured yarns for warp application in weaving are air intermingled to reduce filament snagging in the off-wind of the beams and during the shedding mechanism in weaving. Low, virtually non-visible, air intermingle is often applied to fine filament, low d.p.f. yarns to facilitate

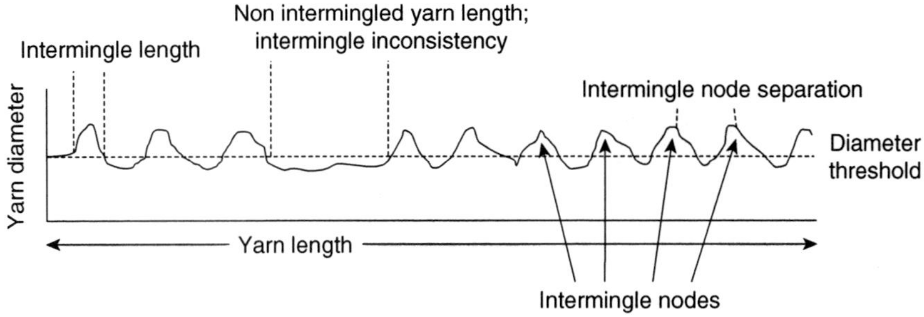

7.2 Intermingle characteristics along a textured yarn.

package off-wind and to reduce tension fluctuations in the downstream knitting or conventional elastane covering operations.

Intermingle characteristics, based on yarn diameter variations due to filament entanglement nodes, are described in Fig. 7.2. Testing equipment is available for measuring intermingle frequency and the resilience of the nodes at pre-selected yarn tensions, so that the stability of the intermingle, when exposed to downstream tensions, can be assessed (see Chapter 10, Section 10.3).

High precision is require in the manufacture of air jets, so that consistency of intermingle is attained across the positions of the draw texturing machine. The manufacturing parameters that affect the process are largely:

- Yarn guides.
- Yarn channel.
- Air inlet.

Yarn guide factors include the positioning of the guides (and yarn angles) at entry to and exit from the jet. Yarn channel factors include:

- Cross-sectional shape.
- Cross-sectional area.
- Length.
- Surface finish of yarn contact areas.
- Longitudinal form.

Air inlet factors include:

- Cross-sectional area.
- Angle in relation to yarn channel.
- Cross-sectional shape.

Modern air jet designs almost invariably feature an open–close system for easy yarn threading and an automatic cut-off of the compressed air feed to the jet when open for threading. The cross-section shape of the yarn chamber

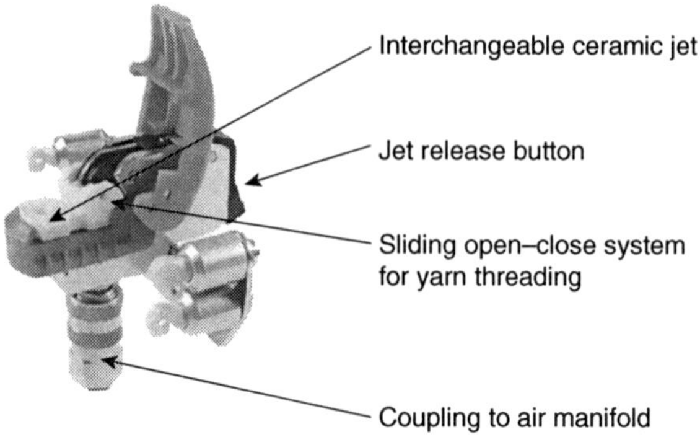

7.3 Heberlein SlideJet, HFP15-2. (Photo © Oerlikon Textile Components.)

is commonly a D-form, where the air flow impacts on a diametrically opposing ceramic flat surface. This is a design feature utilised by Heberlein in their widely-used SlideJet, where the ceramic plate slides open for yarn threading and closes for the intermingling operation (Fig. 7.3).[1] Yarn channel and air intake dimensions to the air jets differ for optimum performance to be achieved over the range of yarn linear densities that are typically texturised by the false twist route. Today, ceramic jet inserts can be interchanged on standard jet support housings to accommodate this need.

In some processes, polymer debris (snow) is carried forward by the yarn from the texturing zone. As a result, air jets need to be periodically cleaned with a brush to maintain intermingle performance consistency. Also in air jet design, it is important that the open–close mechanism does not allow any build-up of deposits, which can affect the geometry of the ceramic plate diametrically opposite the air intake to the jet. The open–close mechanism of the SlideJet is self-cleaning during the slide action and so eliminates this potential problem.

Optimising the parameters in the intermingling process is important for desired intermingle properties to be achieved. Parameters that affect, and should be considered when optimising, intermingle performance are:

- Process specification.
- Yarn characteristics.

Process specification factors are:

- Air pressure.
- Yarn speed.
- Yarn tension.

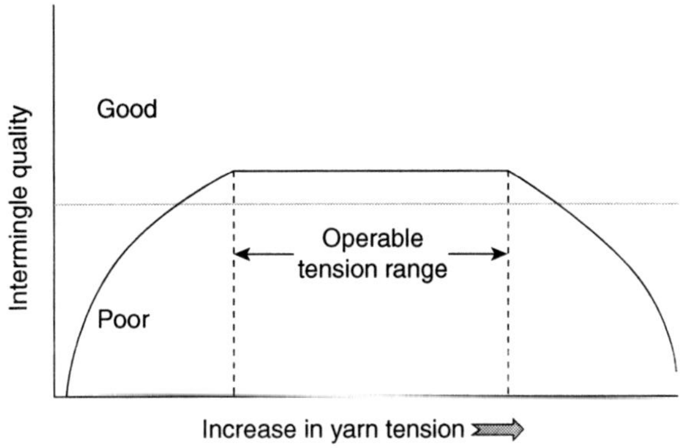

7.4 Effect of yarn tension on intermingle quality.

Factors in yarn characteristics are:

- Linear density (dtex).
- Filament fineness.
- Filament cross-section.
- Yarn surface properties (lubricant, surface debris).

Yarn tension in the intermingling zone is important. When the tension is too high, more energy is required to entangle the filaments via higher air pressure, and often good intermingle consistency is not achieved. When the tension is too low, yarn friction at the jet tends to cause upstream yarn flutter and tension variation, and intermingle consistency is poor. It is therefore important to apply pre-determined yarn tension in the intermingling zone for stable conditions across the texturing machine to be realised (Fig. 7.4).

Air pressure affects intermingle consistency and hence the number of entanglements per metre (Figures 7.5 and 7.6). The stability of the entanglements under tension applied to the yarn also increases with increase in air pressure. Increasing yarn speeds in the draw texturing process necessitates higher air pressures for good intermingling consistency. There are speed thresholds for individual yarns, beyond which consistency is not satisfactory.

For general guidelines, the effect of air jet process parameters on intermingle properties are summarised in Table 7.1. For the purpose of air intermingling draw textured yarns, air jets can be located in two alternative places in the thread path of the draw texturing machine:

(i) Between intermediate feed and second heater entry (or take-up if no secondary heater), which is the preferred option and currently used in the industry.

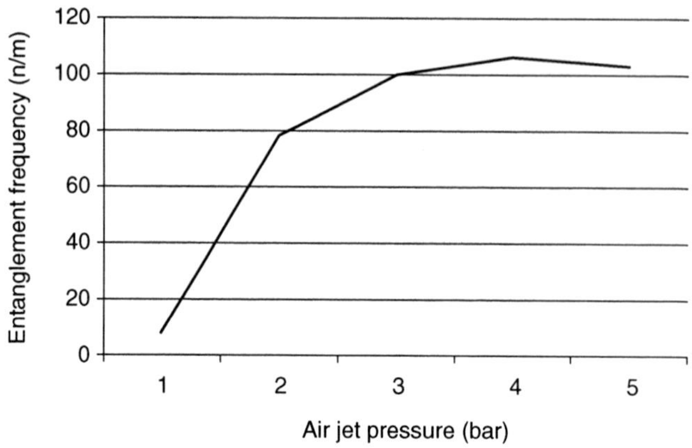

7.5 Example: Effect of air pressure on node frequency, two-fold 167 dtex f34 polyester.

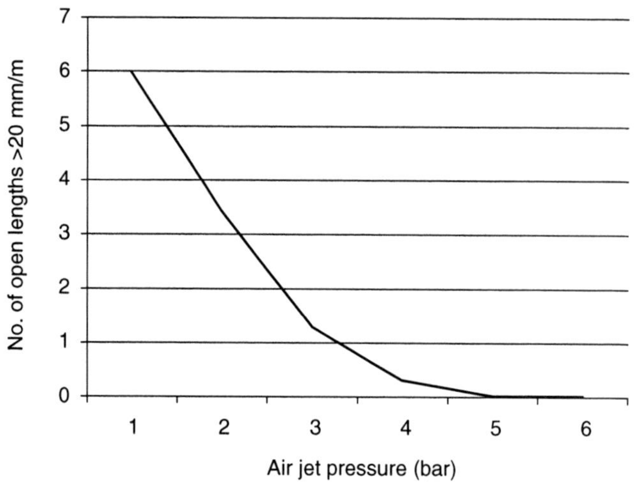

7.6 Example: Effect of air pressure on number of open, non-intermingled lengths, two-fold 167 dtex f34 polyester.

(ii) Between secondary heater outlet (or intermediate feed if no secondary heater) and take-up.

The advantages of the first option are:

- Better jet accessibility.
- Ease of yarn combination for multi-fold operation.
- Constant tension (not influenced by take-up traverse).
- Nodes can be better 'fixed' through light thermal treatment in the second heater (preferred for warp application).

Table 7.1 Guidelines: Air jet parameter effects on draw textured yarn properties

Yarn description (filament fineness)		Node frequency	Node strength	Open lengths	Yarn decitex	Yarn tenacity	Yarn elongation	Crimp value	Crimp stability
> approx. 1.5 d.p.f.	Air pressure	↗	↗	↘	↗	↘	↘	↘	↘
	Yarn tension	⌒	⌒	⌒	⌒	‿	‿	‿	‿
	Process speed	↘	↘	↗	↘	↗	↗	↘	↘
< approx. 1.0 d.p.f.	Air pressure	↗	↗	↘	↗	↗	↗	↘	↘
	Yarn tension	⌒	⌒	⌒	⌒	⌒	⌒	‿	‿
	Process speed	↘	↘	↗	↘	↘	↘	↘	↘

Note: upward arrows indicate that increasing the air pressure, yarn tension or process speed increases the relevant parameter.
downward arrows: vice versa.
Curves indicate that the parameter passes through a maxima or minima.

7.7 Heberlein air jets operating on a Barmag draw texturing machine fitted with additional feed shaft for yarn tension control.

- Easy accessibility to incorporate extra yarn feed shaft for optimum tension control.

The second option has the advantage of higher filament temperatures with easier intermingling, but jet accessibility is poor and without the use of extra yarn feed shaft, tension peaking from wind traverse causes node inconsistency.

Figure 7.7 shows Heberlein air jets operating in an interlacing zone of a Barmag FK6 draw texturing machine with an additional feed shaft for yarn tension control.

There are two other positions for air jets in the draw texturing process, used to achieve other properties:

(i) Pre-primary heater to facilitate filament migration.
(ii) Immediately after second heater (de-torque jet).

Characteristics of the first application include the following:

- Ideally incorporating extra yarn feed shaft for tension control, otherwise simply fitted within the creel thread path prior to the input feed.
- Low air pressure, approximately 0.4 bar, is applied for micro filament application to reduce broken filaments and tight spots, and facilitate downstream processing. Use of air jets in this location facilitates filament migration during twisting in the texturing zone and reduces the exposure of individual filaments to breakage. Moreover, the 'compacting' of the resultant textured yarn significantly reduces any tendency to filament snagging during off-wind of the textured packages in downstream processing.[2]

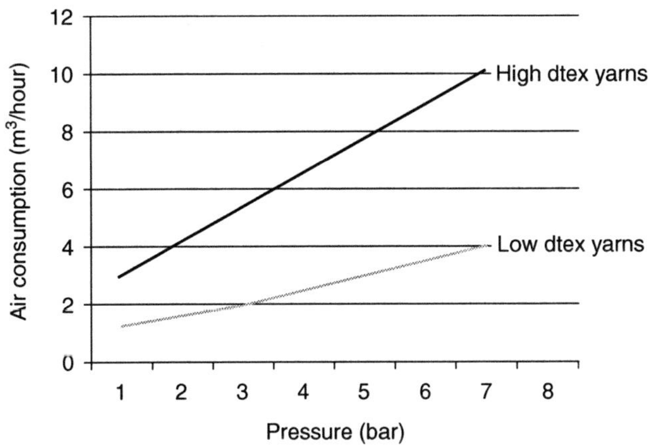

7.8 Typical air consumption range of jets at various air pressures.

- This technology can also be used to reduce the 'floating filament' effect associated with seven-filament polyamide draw textured yarns (See Chapter 8, Section 8.3).

Characteristics of the second application include reducing residual torque in twin-heater yarns, through imposing a light 'false twist effect' in the second heater, thus reducing or eliminating twist liveliness for better downstream yarn performance.[3]

Air jets are commonly used in the draw texturing process but air consumption adds significantly to the overall process costs of the yarn. Air consumption is influenced by the design of the air jet so it is important in intermingling operations to select the best air jet or insert for achieving the required intermingling properties at the lowest possible air consumption. Figure 7.8 shows a typical air consumption range for intermingling yarns from around 11 dtex to 440 dtex using different air jet designs. The effects of air consumption on yarn manufacturing costs are demonstrated, using a 78 dtex draw textured yarn process as an example, in Fig. 7.9.

7.3 Downstream process issues

The air jet design largely determines node frequency. Ideally nodes should not be visible in the fabric and should be released under the tensions experienced in the downstream process. In practice this is not easy to achieve:

- Reducing air pressure reduces node strength, easing node release under tension but increases node-to-node irregularity. Nodes can remain randomly in the yarn as a result, creating 'fleck' appearance in a fabric. Weave structures may be limited by such residual node effects. Careful

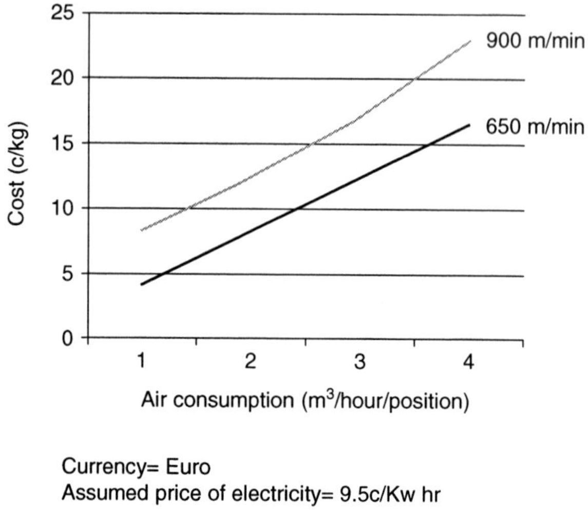

7.9 Indication of air costs (eurocents) for intermingling a 78 dtex product in the draw texturing process.

process optimisation in the intermingling zone is necessary to eliminate such shortfalls of the process.

- For combination yarns, such as marle effect yarns, node stability and frequency must be of the highest quality, otherwise the desired regular marle effect in fabric is offset by random stripes.
- High intermingle node strength and frequency tend to make fabric handle harsher than use of the more expensive twisted yarn, but the relatively lower process cost of air intermingling makes it a first choice.
- High intermingle node strength and frequency reduce tendency to fabric 'pilling'.

7.4 References

1 www.components-oerlikontextile.com

2 C. Atkinson, J. Spahlinger *et al.*, Temcooler: Direct Active Yarn Cooling in Draw Texturing, *Chemical Fibers International*, 54, 336, Oct. (2004).

3 N. Sears, *Premium DTY Air Jet Performance, D-TORK JET: Leading Edge Technology for High Efficiency De-torqueing*, IMS Inc. Winston-Salem, USA.

8
Optimisation of process parameters in yarn texturing

Abstract: This chapter reviews ways of optimising process parameters in yarn texturing. It discusses the influence of texturing parameters on yarn characteristics. It then reviews process control from single-position to plant-scale operation.

Key words: false twist texturing, process parameters and control.

8.1 Influence of texturing parameters on yarn characteristics and performance

The draw texturing process is influenced by a number of process parameters. Process optimisation is aimed at satisfying product requirements for the yarn end uses and establishing a safe operating window, which ensures good process performance and yarn quality consistency across all positions of the draw texturing machine. The raw material POY has to be taken into account. In many cases, texturing operations operate independently and purchase their POY. As a result, they have no opportunity to fine-tune spinning conditions to suite their draw texturing process. In such instances, operating windows in the draw texturing process often have to be compromised, mainly taking into account process speed, yarn linear density and yarn elongation.

The parameters available for optimising the draw texturing process can be segregated into three zones, as shown in Table 8.1. Of these zones, the primary texturing zone is the most important, as this largely sets the basis for yarn tensile performance, crimp value, yarn linear density and process performance in terms of process breaks and quality faults. A summary outlining the effect of primary zone parameters on draw texturing process performance is shown in Table 8.2. The key parameters in the primary zone are draw ratio and selection of friction disc type and combination, all of which affect the maximum speed attainable with respect to the onset of process instability (example, Fig. 8.1). Choice of draw ratio is also important,

Table 8.1 Parameters available optimising the draw texturing process

Primary zone	Secondary zone	Take-up zone
Process speed	Yarn overfeed	Take-up overfeed
Draw ratio	Heater temperature	Wind angle
Heater temperature	Air intermingling	Edge disturbance
Applied **twist**		Traverse length
		Lubricant level

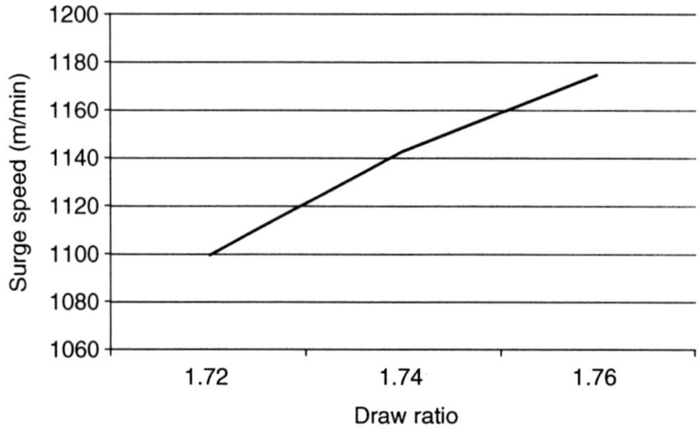

8.1 Effect of draw ratio on process instability using 9 mm diameter PUR friction discs in 1-5-1 combination on polyester 167 dtex f34.

as this determines the yarn linear density and tensile performance (examples, Figures 8.2 and 8.3). Together with disc combination and heater temperature, draw ratio also determines the yarn crimp characteristics (example, Fig. 8.4). Regarding the draw texturing of polypropylene, the yarn tensile properties tend to be especially sensitive to process speed and heater temperature (Table 8.3 and Fig. 8.5).

The secondary zone is only of interest if a stabilising secondary heater is being used for processing draw textured set yarns. Set yarns are usually applicable to polyester processing, but are also occasionally used for polyamide yarns, where specific, lower crimp values and fabric appearance are required. Clearly, the secondary zone, when the machine is fitted with secondary heater and the set heater is in use, has no effect on primary zone process tensions, process instability speed and short term tension transient faults originating from the primary zone or inherent from POY faults. Similarly, the yarn tensile properties are predominantly governed by primary zone process settings. The secondary zone predominantly affects yarn crimp values and residual torque. The influence of secondary zone effects on process performance is outlined in Table 8.4. An example of the effect of

Table 8.2 Primary texturing zone effect on yarn characteristics

Increase in:	Effect on: Tension, T1	Tension T2	Yarn decitex	Tenacity	Elongation	Crimp value	Crimp stability	Filament loops	Broken filaments	Yarn breaks	Residual torque	Surge speed	Tension transients	Dye shade variation	Dye shade (lighter)	Fabric smoothness
Process speed	↑	↑	↔	(↓)	(↓)	↓	↓	↑	↑	↑	↑	↔	↑	↑	↓	↓
Draw ratio	↑	↑	↓	∩↓	↓	(↑)	↑	↔	↑	↑	↔	↑	∪↑	∪↑	↑	↑
D/Y	↔	↓	↔	↔	↔	↔	↔	↔	↓	↔	↔	(↑)	↔	↔	(↑)	(↑)
Heater temperature	↓	↓	↔	┐↓	┐↓	↑	↑	↔	↑	↑	(↓)	(↓)	(↑)	∪↑	↑	(↑)
Twist (no. of discs)	↓	↓	(↓)	(↓)	(↓)	↑	↔	↓	↓	∪↑	↓	↓	└↑	∪↑	(↑)	↑

Key:
↑ ↓ = significant effect in direction of arrow
(↑) (↓) = slight effect in direction of arrow
↔ = no (or very little) effect

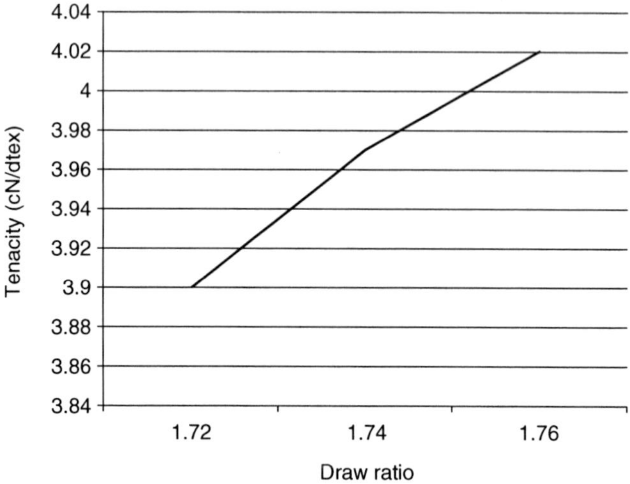

8.2 Effect of draw ratio on yarn tenacity using 9 mm diameter PUR friction discs in 1-5-1 combination on polyester 167 dtex f34 at 900 m/min.

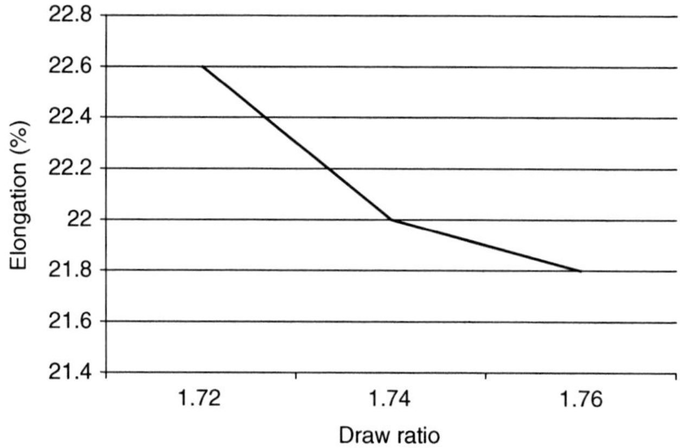

8.3 Effect of draw ratio on yarn elongation using 9 mm diameter PUR friction discs, 1-5-1 combination on polyester 167 dtex f34 at 900 m/min.

secondary heater temperature on yarn crimp characteristics is shown in Fig. 8.6.

The take-up zone parameters are predominantly used to optimise package build for good off-wind performance in downstream processing, i.e. for no yarn snagging and breaks, and low off-wind tension peaks. In simple terms, package density is targeted as high as possible, together with a low wind angle for good off-wind performance, but within the constraints of any adverse effect on the yarn. A limited package storage time can also be

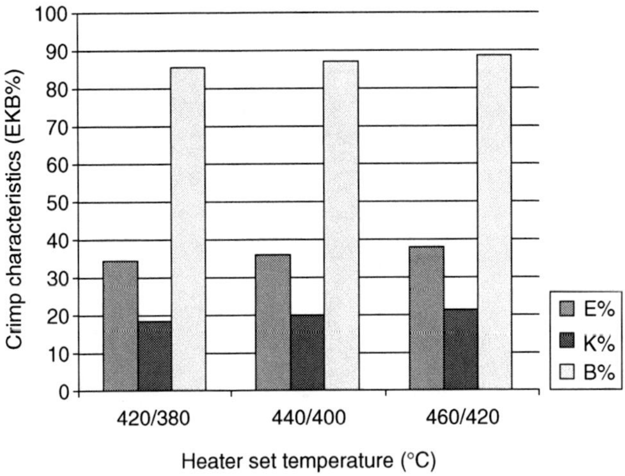

8.4 Influence of HT heater temperature setting on yarn crimp characteristics on polyester 167 dtex f34 at 1050 m/min process speed.

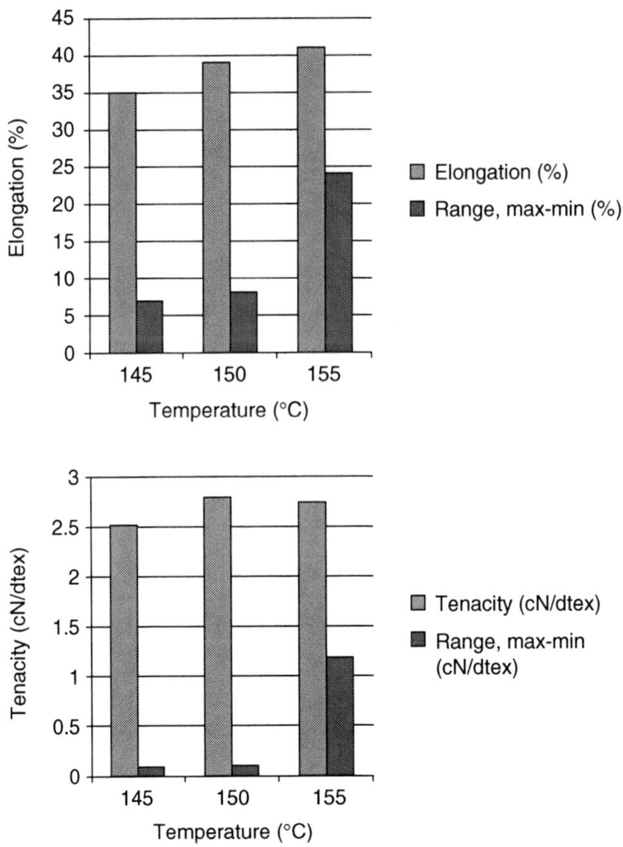

8.5 Effect of heater temperature on tensile performance, polypropylene 220 dtex f40.

Table 8.3 General guidelines for effect of texturing zone parameters on polypropylene yarn

Parameter change	Influence on				
	T1	T2	Tenacity	Elongation	Yarn breaks
Heater temperature (Increase)	⇔	⇔	↷	↗	⬏
Number of discs (Reduce)	⇗	⇧	↗	↑	↓
Intermingle strength (Increase)	⇔	⇔	↗	↗	⇔
Process speed (Increase)	↗	↗	↓	↓	↑
Draw ratio (increase)	↑	↑	↷	↓	↑
DY (Increase)	↗	↓	↘	↘	↗

Key:
↑↓ = significant effect in direction of arrow
⇧⇩ = slight effect in direction of arrow
⇔ = no (or very little) effect

Table 8.4 Secondary zone effects on yarn characteristics

	T1	T2	Titre	Tenacity	Elongation	Crimp value	Crimp stability	Loops	Filament breaks	Yarn breaks	Dye shade Variation	Residual torque	Surge speed	Tension transients	Dye (lighter)	Fabric smoothness
Overfeed ↑	–	–	–	–	–	↓	⇨	⇨	⇨	⇨	↻	↻	–	–	⇧	⇧
Heater temperature ↑	–	–	–	–	–	↓	⇨	⇨	⤴	⇨	⤴	⤴	–	–	⇧	⇧
Air intermingle ↑	–	–	–	–	–	↓	⤵	⇨	➚	⇨	⇨	⇨	–	–	⇨	↓

Key:
↑↓ = significant effect in direction of arrow
⇧⇩ = slight effect in direction of arrow
⇔ = no (or very little) effect

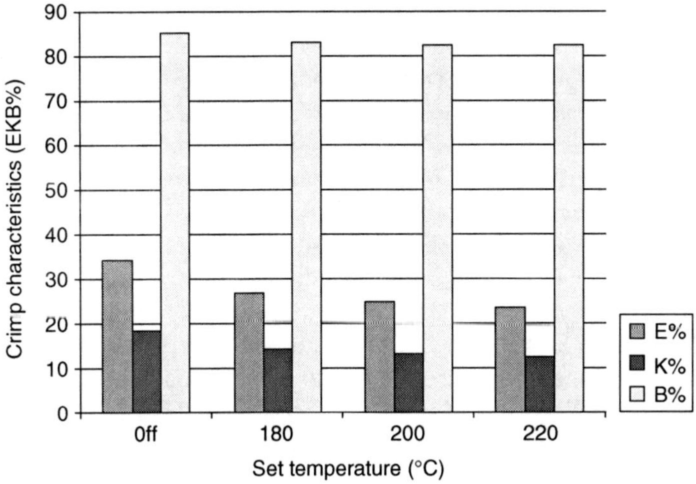

8.6 Influence of secondary heater on crimp characteristics on polyester 167 dtex f34 at 1050 m/min process speed.

Table 8.5 Take-up zone and package storage effects on yarn characteristics

Parameter	Effect on yarn
Increased wind tension/package density	Reduced residual torque
Increased wind angle/wind tension	Fine filament yarns: *broken filaments at traverse reversal* *short-term crimp and dye defects*
Increased coning oil level	Increased yarn linear density
Increased package storage time over several days	Reduced residual torque
	Better dye shade consistency

beneficial for certain yarns and applications for reduced residual torque in the textured yarn, and in some instances reduced dye uptake variation between yarn packages. Take-up zone parameters have virtually no effect on yarn characteristics, the exceptions being highlighted in Table 8.5.

8.2 Single-position to plant-scale operation

When optimising a draw texturing specification, it is important to consider what machine types are at disposal or indeed what properties are required for the target end use. For example, is the specification aimed at achieving uppermost speed or quality improvement? Also, what is the texturing zone heat–cool profile? The first stage of process optimisation is to select the friction disc type and combination, although it is better, where possible, to standardise disc type across the plant to minimise any complexity. Examples of typical disc types and combinations are shown in Tables 8.6, 8.7 and 8.8 for use on polyester, polyamide and polypropylene yarns, with 37 mm or 36 mm centre spacing for the twist unit spindles. For folded heat–cool zones, it is preferable to apply a relatively high number of working discs so that adequate torque is applied for a sufficiently high twist level to be realised within the heater thread path.

The second stage of process optimisation is to examine trends in draw ratio for the selected friction disc combination over a realistic range at a conservative process speed, in order to determine trends in yarn linear density, tenacity and elongation. Where both S and Z twist components are required, this fundamental exercise should be carried out on a few texturing positions for each twist direction. It is beneficial at this stage to graph trends in yarn tensions, linear density (decitex) and tensile properties as a basis for future reference.

For the draw ratio conditions that bring about the desired yarn linear density and tensile properties, yarn crimp properties should be determined and corrections made, where necessary, through alternative disc combination and heater temperatures. As a rule, increasing disc numbers and increase in heater temperature, increase crimp values, but for hard disc systems, loss in tensile performance can sometimes be expected through increase in disc numbers and this must be taken into account.

When the basic specification has been arrived at, it is good practise to determine the process instability speed, simply by increasing the speed for the set conditions in 50 m/min increments and monitoring the process tension at each speed for say 5–10 minutes to establish at what speed tension surging occurs. For process stability assurance, it is advisable to finally set the process speed at least 100 m/min under the speed at which the onset of surging occurs. The above sequence of events in the optimisation of the process specification is summarised in Table 8.9.

Having established a basic process specification, it is then recommended to check its performance for consistency and to obtain an indication of process break rate. This should be carried out on 12 to 24 positions for each twist direction, producing full draw textured yarn packages for 1 to 3 doffs. Attention should be given to the following performance indicators:

- Process tension variation within and between positions (using on-line tension monitoring).
- Mean values and variation in tensile and crimp values.
- Visual inspection for broken filaments and 'tight spots'.
- Package density.
- Process break rate.
- Knit/dye consistency between positions using single-feed circular knit testing.

Table 8.6 Examples of friction disc combinations used on polyester yarns

Yarn count (dtex)	d.p.f	Disc combination	Input guide disc				Working disc				Exit guide disc			
			Diam.	Width	Ra	Material	Diam.	Width	Ra	Material	Diam.	Width	Ra	Material
78	2.3-2.3	1-5-1	52 or 46	9	0.3	ceramic	52	9	–	PUR	52	9	0.3	ceramic
		1-5-K	52 or 46	9	0.3	ceramic	52	9	–	PUR	52	9	0.3	ceramic
		1-7-1	52 or 46	9	0.3	ceramic	52	9	0.6	ceramic	52	9	0.3	ceramic
110	0.5-1.0	1-4-K	52 or 46	9	0.3	ceramic	52	9	–	PUR	52	9	0.3	ceramic
		1-5-K	52 or 46	9	0.3	ceramic	52	9	–	PUR	52	9	0.3	ceramic
	3.3	1-5-1	52 or 46	9	0.3	ceramic	52	9	–	PUR	52	9	0.3	ceramic
		1-5-K	52 or 46	9	0.3	ceramic	52	9	–	PUR	52	9	0.3	ceramic
		1-6-1	52 or 46	9	0.3	ceramic	52	9	–	PUR	52	9	0.3	ceramic
		1-6-K	52 or 46	9	0.3	ceramic	52	9	–	PUR	52	9	0.3	ceramic
		1-7-1	52 or 46	9	0.3	ceramic	52	9	0.6	ceramic	52	9	0.3	ceramic
		1-7-K	52 or 46	9	0.3	ceramic	52	9	0.6	ceramic	49	9	0.3	ceramic
167	1.7-5.2	1-6-1	52 or 46	9	0.3	ceramic	52	9	–	PUR	52	9	0.3	ceramic
		1-6-K	52 or 46	9	0.3	ceramic	52	9	–	PUR	52	9	0.3	ceramic
		1-7-1	52 or 46	9	0.3	ceramic	52	9	0.6	ceramic	52	9	0.3	ceramic
		1-7-K	52 or 46	9	0.3	ceramic	52	9	0.6	ceramic	52	9	0.3	ceramic
		1-8-1	52 or 46	9	0.3	ceramic	52	9	0.6	ceramic	52	9	0.3	ceramic
		1-8-K	52 or 46	9	0.3	ceramic	52	9	0.6	ceramic	52	9	0.3	ceramic
	0.5-1.0	1-5-K	52 or 46	9	0.3	ceramic	52	9	–	PUR	52	9	0.3	ceramic
		1-6-K	52 or 46	9	0.3	ceramic	52	9	–	PUR	52	9	0.3	ceramic
		1-5-K	52 or 46	9	0.3	ceramic	52	9	0.6	ceramic	52	9	0.3	ceramic
		1-6-K	52 or 46	9	0.3	ceramic	52	9	0.6	ceramic	52	9	0.3	ceramic

Table 8.7 Examples of friction disc combinations used on polyamide yarns (nylon 6 and 66)

Yarn count (dtex)	d.p.f	Disc combination	Input guide disc				Working disc				Exit guide disc			
			Diam.	Width	Ra	Material	Diam.	Width	Ra	Material	Diam.	Width	Ra	Material
11	1.6	1-4-1	49.6	6	0.3	ceramic	49.6	6	0.85	ceramic	49.6	6	0.3	ceramic
17	3.4	1-4-1	49.6	6	0.3	ceramic	49.6	6	0.85	ceramic	49.6	6	0.3	ceramic
22	3.1	1-4-1	49.6	6	0.3	ceramic	49.6	6	0.85	ceramic	49.6	6	0.3	ceramic
		1-5-1	49.6	6	0.3	ceramic	49.6	6	0.85	ceramic	49.6	6	0.3	ceramic
22	1.6	1-5-1	49.6	6	0.3	ceramic	49.6	6	0.85	ceramic	49.6	6	0.3	ceramic
		1-5-K	52 or 46	9	0.3	ceramic	52	9	1.1	ceramic	49	9	0.3	ceramic
33	3.3	1-5-1	49.6	6	0.3	ceramic	49.6	6	0.85	ceramic	49.6	6	0.3	ceramic
		1-6-1	49.6	6	0.3	ceramic	49.6	6	0.85	ceramic	49.6	6	0.3	ceramic
44	3.4	1-5-1	49.6	6	0.3	ceramic	49.6	6	0.85	ceramic	49.6	6	0.3	ceramic
		1-6-1	49.6	6	0.3	ceramic	49.6	6	0.85	ceramic	49.6	6	0.3	ceramic
44	1.3	1-5-1	49.6	6	0.3	ceramic	49.6	6	0.85	ceramic	49.6	6	0.3	ceramic
		1-6-1	49.6	6	0.3	ceramic	49.6	6	0.85	ceramic	49.6	6	0.3	ceramic
		1-5-K	52 or 46	9	0.3	ceramic	52	9	1.1	ceramic	49	9	0.3	ceramic
78	1.1	1-6-1	49.6	6	0.3	ceramic	49.6	6	0.85	ceramic	49.6	6	0.3	ceramic
		1-7-1	49.6	6	0.3	ceramic	49.6	6	0.85	ceramic	49.6	6	0.3	ceramic
		1-7-K	52 or 46	9	0.3	ceramic	52	9	1.1	ceramic	49	9	0.3	ceramic
78	2.3-3.3	1-6-1	49.6	6	0.3	ceramic	49.6	6	0.85	ceramic	49.6	6	0.3	ceramic
		1-7-1	49.6	6	0.3	ceramic	49.6	6	0.85	ceramic	49.6	6	0.3	ceramic
		1-7-1	49.6	6	0.3	ceramic	49.6	6	0.85	ceramic	49.6	6	0.3	ceramic
		1-7-1	52	9	0.3	ceramic	52	9	1.1	ceramic	52	9	0.3	ceramic
		1-7-K	52 or 46	9	0.3	ceramic	52	9	1.1	ceramic	49	9	0.3	ceramic
110	3.3	1-7-1	49.6	6	0.3	ceramic	49.6	6	0.85	ceramic	49.6	6	0.3	ceramic
		1-8-1	49.6	6	0.3	ceramic	49.6	6	0.85	ceramic	49.6	6	0.3	ceramic
		1-7-1	52	9	0.3	ceramic	52	9	1.1	ceramic	52	9	0.3	ceramic
		1-7-K	52	9	0.3	ceramic	52	9	1.1	ceramic	49	9	0.3	ceramic
		1-7-K	46	9	0.3	ceramic	52	9	1.1	ceramic	49	9	0.3	ceramic
		1-8-1	52	9	0.3	ceramic	52	9	1.1	ceramic	49	9	0.3	ceramic
		1-8-K	52 or 46	9	0.3	ceramic	52	9	1.1	ceramic	49	9	0.3	ceramic

Table 8.8 Examples of friction disc combinations used on coloured polypropylene yarns

Yarn count (dtex)	d.p.f	Disc combination	Input guide disc				Working disc				Exit guide disc			
			Diam.	Width	Ra	Material	Diam.	Width	Ra	Material	Diam.	Width	Ra	Material
220	5	1-8-1	52	9	0.3	ceramic	52	9	1.1	ceramic	52	9	0.3	ceramic
		1-8-K	52	9	0.3	ceramic	52	9	1.1	ceramic	49	9	0.3	ceramic
		1-8-K	46	9	0.3	ceramic	52	9	1.1	ceramic	49	9	0.3	ceramic
110	1.6	1-7-1	52	9	0.3	ceramic	52	9	1.1	ceramic	52	9	0.3	ceramic
		1-7-K	52	9	0.3	ceramic	52	9	1.1	ceramic	49	9	0.3	ceramic
		1-7-K	46	9	0.3	ceramic	52	9	1.1	ceramic	49	9	0.3	ceramic
78	1.3	1-6-1	52	9	0.3	ceramic	52	9	1.1	ceramic	52	9	0.3	ceramic
		1-6-K	52	9	0.3	ceramic	52	9	1.1	ceramic	49	9	0.3	ceramic
		1-6-K	46	9	0.3	ceramic	52	9	1.1	ceramic	49	9	0.3	ceramic

Table 8.9 Sequence of events for preliminary stages in process optimisation

	Parameter adjustment and settings	Comments
Stages		
	Select disc type and combination, based on speed and crimp value requirements	
	Set heater temperature	Typically between 190–220 °C for PET and PA66, 170–190 °C for PA6
Six positions per twist	Select draw ratio based on decitex and yarn tensile properties (Apply D/Y ratio for a T2/T1 ratio of around 0.8–1.1)	Typically 1.68–1.76 for PET, 1.23–1.34 for PA66 and 1.20–1.22 for PA6 Determine trends for yarn tensile properties with change in draw ratio
	Determine process instability speed for selected conditions	Set speed at least 100 m/min below surge speed
	Where necessary revisit disc combination and heater temperature for satisfying crimp requirements	Test yarn: linear density, tensile properties, crimp value, knit dye in single panel Visually check yarn lengths for crimp tight spots, broken filaments on package taper wall
	If intermingle required, optimise intermingling zone overfeed and air jet pressure for desired entanglement properties	
	If secondary heater required for set yarn, examine range of set heater temperatures on crimp values and select temperature for desired crimp	Adjust overfeed to maintain virtually zero yarn tension at heater entry for each individual temperature setting
	Set take-up specification, building full packages for density measurement and adjust oil roller for required coning oil level	Aim for low wind angle and set overfeed to provide desired package density Adjust edge disturbance setting for no webbing and good package shape at traverse reversal

If quality or process problems are evident at this stage, the process specification needs to be revisited. If there are no problems evident, where possible, the trial quantity should be evaluated in downstream processing before moving into full-scale production within the draw texturing plant. When introducing a new product to production machines, it is also important that the machine and heaters are fully cleaned prior to start-up.

8.3 Resolving process quality issues

There are many common quality issues experienced in draw texturing across the yarn product ranges processed. There are, nevertheless, some quality and process criteria that relate specifically to whether polyester, nylon or polypropylene yarns are being processed and to the yarn count or d.p.f of the yarns. In general terms, however, the solutions to the majority of quality issues can be summarised as shown in Fig. 8.7.

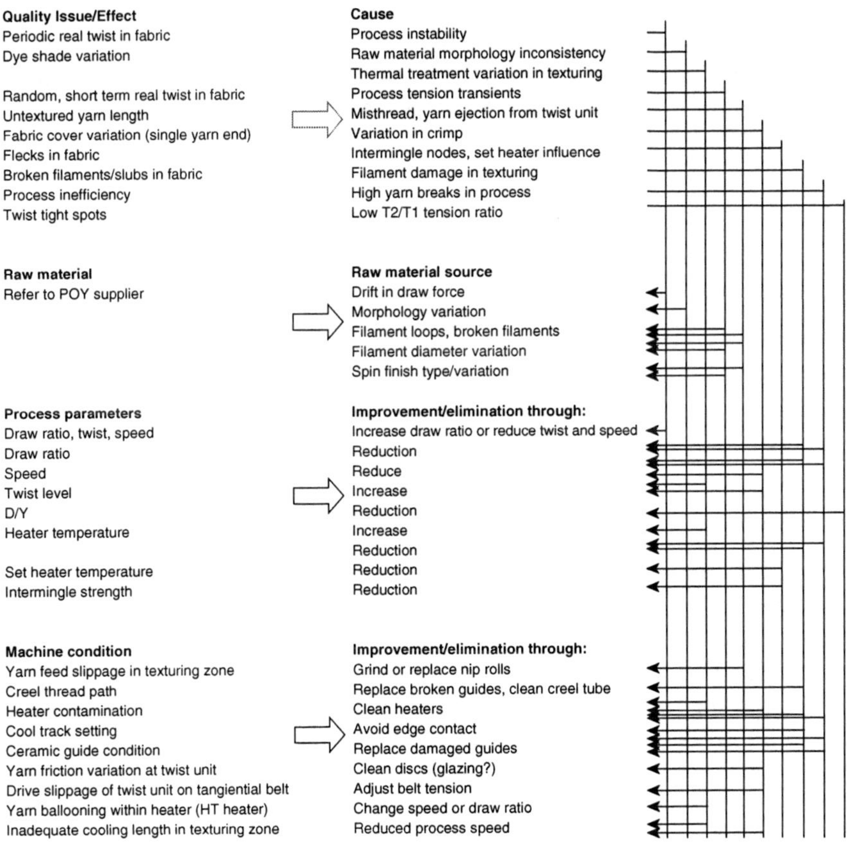

8.7 Quality problems; common causes and solutions.

Where possible, when quality problems are encountered, adjustment to the process specification should be the last resort as a remedial action because this necessitates product lot number change, which is not welcomed by the customers. Often, when process change is necessary, it reflects on the lack of thoroughness and attention to detail given to the process in its initial optimisation stages prior to full production. For the draw texturing of microfilament yarns, common draw texturing inherent faults, e.g. broken filaments, tension transients and tight spots, are shown in Fig. 8.8, together with solutions for eliminating or reducing their occurrence.

For the processing of microfilament yarns, filament migration within the twisted bundle in the texturing zone is difficult due to the high number of filaments, and attempts to impart too high a twist level can lead to broken filaments and tight spots. Consequently, it is advisable to operate with relatively lower friction disc numbers than those used on higher d.p.f yarns of the same yarn linear density. Broken filaments for such yarns can often be reduced by lowering draw ratios and thus process tensions, but care should be taken not to reduce the yarn crimp stability to the extent that tensions in the downstream process adversely affect crimp uniformity in the fabric.

For fine decitex nylon yarns, yarn break rates in the process tend to be a common problem, which can be addressed by:

- Reducing draw ratio.
- Reducing process speed.
- Reducing twist level (although high twist levels are usually required to generate good elasticity in the yarn for four-feed knit hosiery end-use).
- Reducing heater temperature.

However, for these yarn types, machine condition and plant disciplines often contribute to break rates. These include:

- Inadequate heater cleanliness.
- Poor nip roller condition.
- Poor POY package alignment in creel.
- Bad practices in POY handling causing yarn damage at creeling.

Moreover, the spinning of nylon 66 is prone to gel formation in filaments due to inhomogeneity in the melt. This is a common cause of tension transients in the draw texturing process and always leads to short-term textured yarn faults that often cause yarn process breaks.

Concerning fine decitex polyamide yarns, it is worthwhile referring to a common problem associated with the texturing of seven-filament yarns, typically 22 dtex f7 for ladies' hosiery application. Such yarns can be particularly problematic due to excessive lengths of non-textured filaments that can create plucked or drawn threads in downstream processing. This phenomenon can be influenced by the friction characteristics between

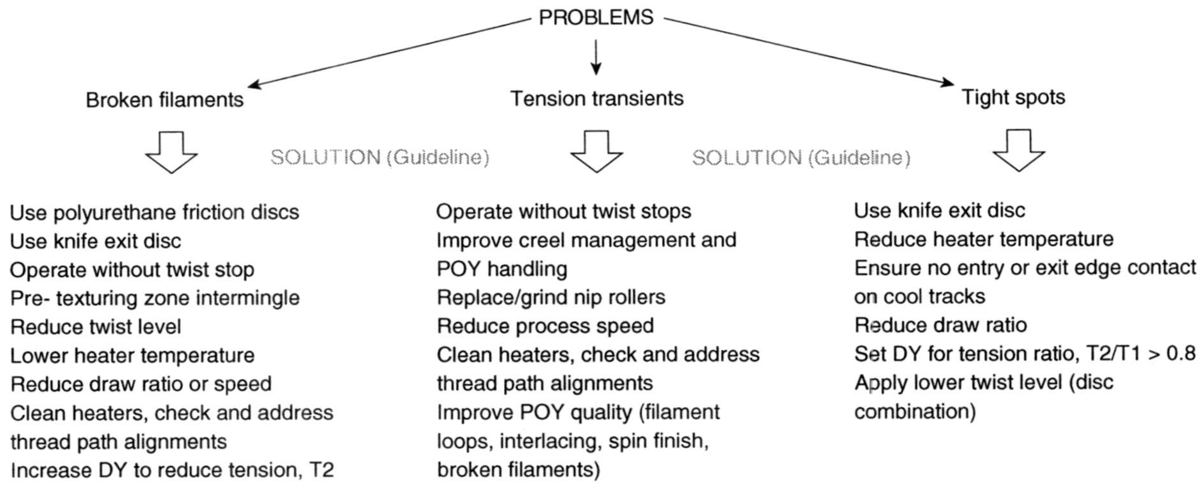

8.8 Common faults and solutions when texturing microfilament yarns.

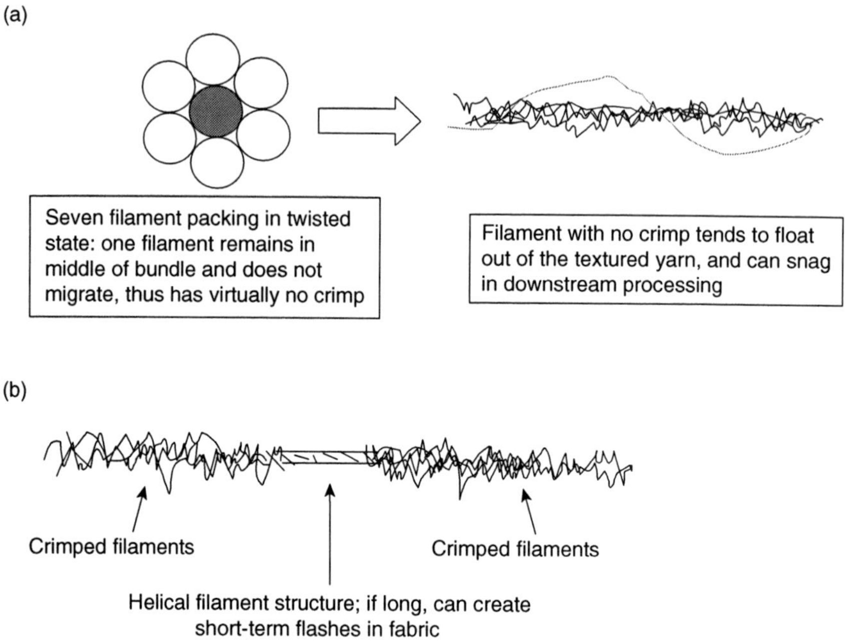

8.9 Common faults associated with draw textured polyamide 22 dtex f7. (a) Floating filaments (b) Corkscrew effect.

filaments due to the spin finish type, twisting torque and tension at the initiation of twist in the thread path. The 'floating filament' effect cannot be totally eliminated (Fig. 8.9) but can usually be reduced by:

- Increasing twist level (leads, however, to process speed limitations).
- Reducing process speed.
- Introducing pre-intermingling at low air pressure to the texturing zone.

The 22 dtex f7 product is also generally prone to 'corkscrew' faults, which can be reduced or eliminated through increase in process tension via draw ratio; here, the maximum tension applied is being limited by the onset of process breaks. Corkscrew faults can also be caused by irregular run back of twist to the heater in the texturing zone, typically brought about by yarn contact on cool track entry or exit edges, as a result of thread path setting indiscipline. Large corkscrew faults can be visible in the knitted fabric.

Spun dyed polyester yarns and coloured polypropylene yarns are particularly prone to broken filaments due to the filament migration difficulties during twist formation in the texturing zone and high friction. This can be addressed by normal means through reduction in draw ratio, process speed, heater temperature or twist level. Such yarns can be abrasive and cause

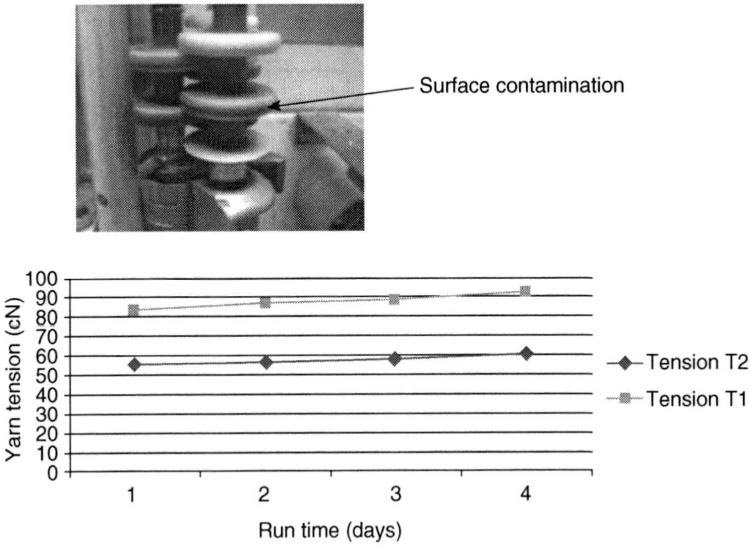

8.10 Rapid increase in tension due to disc glazing with a polypropylene yarn.

rapid wear of contact surfaces in the texturing zone, and contact surfaces such as ceramic guides and cool tracks must be regularly inspected for damage. Moreover, the surface roughness consistency of hard material friction discs can be affected through surface contamination, leading to loss of friction (disc glazing) (Fig. 8.10). This necessitates regular ultrasonic cleaning of the discs as a preventive measure. Disc glazing in the draw texturing process is evident in progressive increase in *T2* and *T1* process tensions. This increase can be rapid, for example, with some spun dyed polypropylene yarns but is more likely to increase slowly over months in the event of glazing incidents on other yarns.

9
Draw textured yarn variants and speciality yarns

Abstract: This chapter reviews draw textured and speciality yarns. It discusses double-density and other machine variants for draw textured speciality yarns.

Key words: false twist texturing, draw textured yarns, speciality yarns.

9.1 Double-density machines

To obtain more cover in the fabric through higher crimp frequency in the filaments, two textured yarns can be combined after the yarn output feed in draw texturing, usually as an S and Z twist combination. This, however, restricts the output of the machine, as only 50% of the take-up positions are utilised. As a consequence, 'double-density' machines were developed, the machines being equipped with extra heating and cooling tracks, double creel capacities for the raw material and twin twist units (Fig. 9.1). This enabled full utilisation of the take-up positions, increasing the machine capacities and making the process economics more viable.

Today, more modern machines can be purchased with texturing zone designs that incorporate a two-fold, double-density capability, which incorporate two single, standard twist units per processing position.[2] The double-density concept also, has additional advantages in draw-texturing in that different raw material yarns can be combined in the process for different aesthetics, handle and colour effects in the fabric. This opportunity is widened by the addition of extra yarn feeds to the machines for enabling different draw ratios or overfeeds to be applied to the relevant processing zone.

9.2 Machine variants for draw textured speciality yarns

Because high volume demands tend to be low for speciality yarns that are processed by the multi-feed, double-density means, the concept tends to

9.1 Example of twin twist unit, TEMCO FTS 4461.[1] (Photo © Oerlikon-Barmag.)

have more application in the smaller, modular build machines with single motor drives, designed for process flexibility. Machine variants for draw textured speciality yarn are summarised in Table 9.1. It is fair to say that there are limited opportunities for speciality product development in the draw texturing process. Most of the development opportunities stem from the combination of existing feeder yarns or the use of novel raw material yarn variants. Examples of such opportunities are listed in Table 9.2; these can be used for single draw textured yarn processes or for the processing of combination yarns for special effects.

The most commonly used routes for draw texturing speciality yarns are:

- Yarn combinations via the double density approach for multi-colour products.
- Yarn combinations for differential filament shrinkage.
- Combination of elastane and draw textured yarn.
- Simple double density application for yarn softness and increased machine output.
- Thick-thin yarns via pre texturing zone intermittent thermal treatment.

Table 9.1 Draw texturing machine variants for speciality yarns

Double density	Twin texturing unit per position Double creels Double density heater tracks Double density cool tracks All take-up positions utilised
Double density, 2- or 3 input feed	Different draw ratios applied to feeder yarns Can combine flat (or even staple) and textured yarn with flat yarn bi-passing heater All take-up positions utilised
Single density, multi-input feed	Yarn combination through same texturing zone thread path All take-up positions utilised
Hot pin in the a pre-texturing zone with extra yarn feed for tension control	Can be used to combine flat with textured yarns or textured yarns with different shrinkage Thick-thin effects by oscillating yarn on a hot contact surface
Extra post-texturing zone feeds	Combination of elastane with textured yarn via air intermingle

Table 9.2 Examples of raw material yarn variants that can be used for special characteristics and effects

Variant	Description	Characteristic
Modified filament cross-section	Trilobal	Brightness, lustre (light reflectance)
	Multi-lobal	Dullness (light diffusion)
	Dogbone	Softness (filament bending)
Modified polymer lustre	Bright	Shiny appearance
	Semi dull	↓
	Full dull	Matt appearance
Filament fineness	Microfilament ≤ 1.0d.p.f.	Softness, fabric cover
Hollow filament yarns		Thermal protection
Modified polymer yarns	Cationic dyeable	Colour differential
	Polybutyl terephthalate, PBT	Incresed elasticity
	Polytrimethylene terephthalate, PTT	Increased elasticity
	Spun dyed	Built-in colour
	Anti-static additive	Reduced static
	Water absorbent additive	Moisture absorbancy
Bicomponent composite filament yarns	Conjugated:	
	side by side	Self crimping
	sheath core	Self crimping
	splitable/multi-layer	Softness

9.2 Combined yarns to provide a simple two-tone, marle effect.

An example of a two-tone marle effect produced by combining two spun dyed Polyester yarns via the double density processing route is shown in Fig. 9.2 in the form of a knitted fabric.

9.3 References

1. *TEMCO Double Unit FTS4461 for Polyester, Polyamide and Polypropylene Yarns*, TEMCO Textilmaschinenkomponenten GmBH, Technical Document TC 12101/1 EA (2001).
2. *MPS As Creative as Market*, Oerlikon Barmag, Brochure OBA 307 e/7 (2007).

10
Process control and quality assurance in yarn texturing

Abstract: This chapter reviews process control and quality assurance in yarn texturing. It discusses in-process controls, laboratory controls and final product inspection as well as statistical process control.

Key words: false twist texturing, process control, quality assurance.

10.1 In-process controls

To ensure consistent and good product quality in draw textured yarn, numerous disciplines are required in the texturing plant, taking into account:

- Raw material quality and consistency.
- Operational disciplines and training.
- Process optimisation.
- Identification and rectification of faulty processing positions.
- Yarn testing for product quality verification.
- Statistical process controls.
- Definition of product grade criteria.
- Monitoring of process performance.
- Product final inspection and packing.
- Package handling.
- Transportation of packages within the plant and to the customers.
- Maintaining adequate records of performance and remedial action needs.
- Performance review, including any customer complaints analyses.

Raw material consistency is paramount for good texturing performance and it is advisable to carry out a thorough assessment of suppliers for assurance that they have sound quality management systems in place. Raw material lot numbers and yarn descriptions should be checked, both on packing labels and on the product description labels used inside the POY tubes, prior and during creeling, for assurance that no errors have been made by the supplier. Also, any raw material yarn showing transit or packing damage should be isolated for consultation with the supplier.

Within the draw texturing plant, the process performance and product quality are controlled and monitored through in-process monitoring, periodic product testing using laboratory testing means, through to visual inspection of all packages at packing. Overall performance is also very much influenced by the process conditions optimised prior to full production start-up, machine condition and operator training and efficiency.

Today, on modern machines, process settings and shaft-speed monitoring are carried out through electronic drive and control systems and rarely are errors made in the setting-up of process specifications. Previously, however, yarn feed speeds in the sub-sections of the machine were set by gear changes and it was necessary to check with a tachometer all shaft speeds of the machine prior to threading the positions at production start-up for assurance that the correct gears had been selected.

Regarding in-process controls, operators must be well trained and disciplined in their work. They must check every position after threading for assurance that all guides are threaded correctly. Procedures must also be in place to check and remove any yarn wraps on twist unit spindles prior to threading, to avoid loose yarn ends or filaments contacting the running thread line within the twist unit.

Online tension monitoring on every process position is now commonly used in the industry as an in-process control means (Fig. 10.1). Yarn tension sensors, fitted within the post-twist unit thread path, measure *T2* tensions to detect and record tension anomalies typically inherent from:

- Raw material faults, largely short-term, caused by filament loops, broken filaments or filament gels in nylon 66 (Fig. 10.2).

10.1 Online tension monitoring sensors, Barmag-Oerlikon 'Unitens'.

- Friction disc issues such as disc damage, progressive tension change due to polyurethane disc wear or disc surface glazing.
- Friction unit drive slippage on the tangential drive belt.
- Operator mis-threads in the texturing zone.
- Yarn feed slippage caused by poor operator disciplines or worn nip rollers in the texturing zone (Fig. 10.3).
- Process tension drift due to change in POY draw force.

For twist units with individual motor drives, tension sensors tend to be fitted to the housing of the twist unit base and do not need separate support rails.

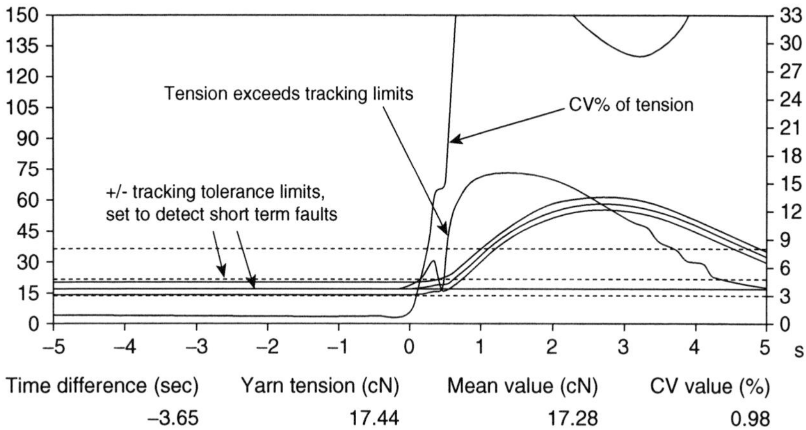

10.2 Example of short-term tension fault, inherent from raw material POY.

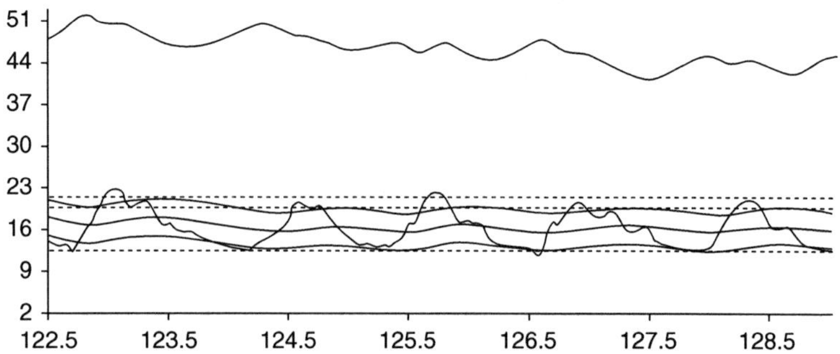

10.3 Loss of draw and process tension due to operator failure to engage the input feed nip roller onto the drive, leading to tension surging. The lower tension tolerance under these twist and speed conditions for the onset of surging are exceeded, causing periodic process instability, as described in Section 5.2.

Tension monitoring provides a means of detecting processing faults where tensions deviate beyond tolerances that can be set for an individual process. Tolerances can be set globally across the machine for dye shade anomalies, operator mis-thread detection and for the detection of short-term faults that can occur within the global tolerance setting on individual positions (Figures 10.2 and 10.4). Usually, the system provides at least the facility for downgrading textured packages according to the duration of the faults that exceed maximum or minimum tensions, tension variation that exceeds a selected coefficient of variation and the total number of faults within a package according to the selected tolerances.[8] In most cases, the system is interfaced with the yarn cutters that are activated by yarn sensors to prevent yarn wraps on feed shafts for yarn break incidents. By so doing, a cut can be activated when a position exceeds selected tolerances for package quality downgrade. LED indicator lights on the tension sensors serve the purpose of informing operators the quality category of a yarn package at doffing or for the cut package.

Tension data are usually stored in online tension monitoring systems and summarised in doff and shift reports for break rate, quality down grade, output efficiency and positional performance analyses. These provide an overview of process performance and identify remedial action needs.

In recent years, a creel management system has also been developed, whereby yarn sensors monitor the tail transfers of the raw material yarn.[1] This system is integrated into the online tension monitoring facility and provides information on tail transfer efficiencies and associated tension faults in the draw texturing process, thus integrating the entire process into a more effective Total Quality Management System.

10.2 At yarn package doff

At package doffing, packages are removed and labelled by the operators for traceability purposes, i.e. date, time and machine position history. Where automatic package doffing applies, yarn breakages during the doff sequence are monitored by the doffing system to identify positions that are problematic and requiring attention. Doffing efficiency reports are also available. In a more recent development, package marker facilities are available via linkage to the online tension monitoring system to enable segregation of A-quality and downgrade packages.[1]

10.3 Laboratory controls

10.3.1 Knit-dye testing

Online tension monitoring, in particular, is good for detecting short-term faults in the yarn and operator mis-threads. It does, however, not always

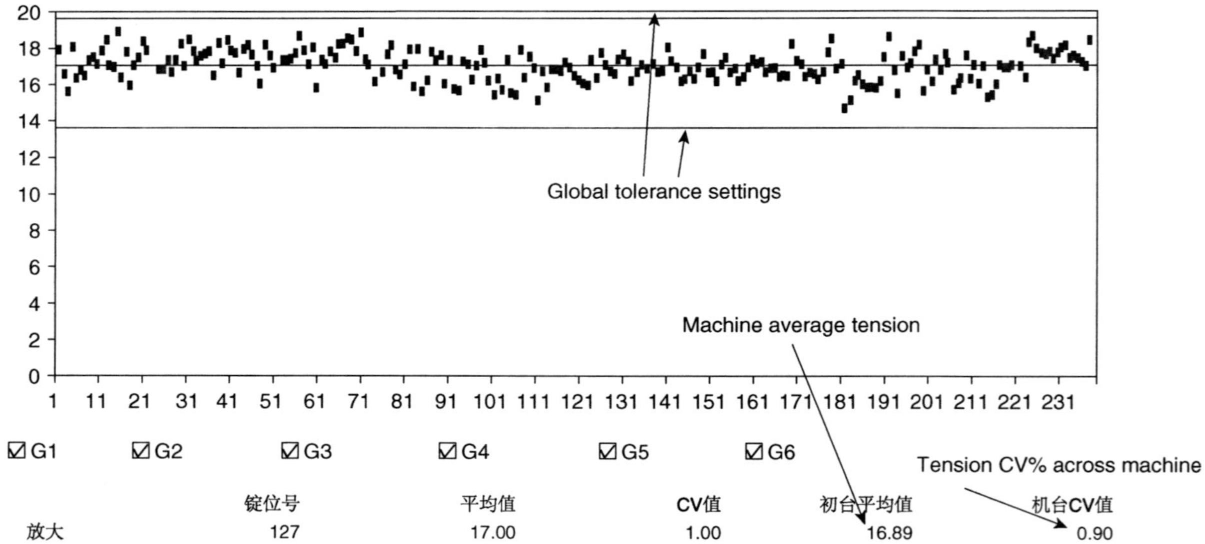

10.4 Real-time process tensions on a machine (Barmag-Oerlikon Unitens).

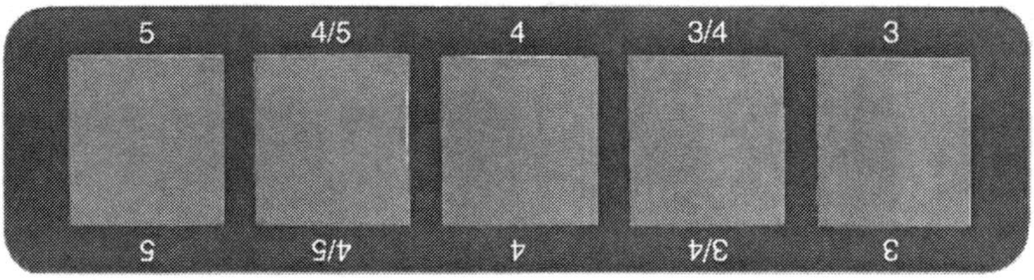

5 = no visual change, 3 = large visual change

10.5 Grey scale for assessing change in dye shade.

detect some subtle longer-term morphological changes in the raw material yarn and slight changes to heat transfer in the texturing zone, both of which can lead to dye shade variation in the product. Therefore, for the more critical yarn end-uses, such as warp application, controlled lengths of the outside layer of all packages are knitted sequentially on single-feed circular knitting machines. The knitted panels are dyed with critical dyestuff to detect and isolate dye shade anomalies that are visually considered under a UV light cabinet to be outside the grey-scale selection criterion (Fig. 10.5).[2] For this visual assessment, a white–black, dual sided plastic board is inserted into the knitted tube providing light tension so that the knitted panel can be freely moved along the board for positional assessment. The white side of the board serves to identify positions that exhibit low cover due to crimp value anomalies through transmission of reflected light.

Alternative and more automated test methods are available to detect dye shade variation based on the thermal shrinkage characteristics of the textured yarn.[3] It is claimed that up to 1000 packages can be tested per eight-hour shift using such equipment, which also incorporates the facilities to test yarn shrinkage and intermingle performance. Nevertheless, knit-dye testing still currently remains the most commonly used testing method for detecting dye shade variation in draw textured packages.

Quality control laboratories also daily measure, through testing sample packages from production machines, the following yarn properties for assurance that tolerances are maintained:

- Yarn tensile performance (linear density, tenacity and elongation).
- Crimp values.
- Intermingle frequency and strength.
- Coning oil level.

10.3.2 Yarn crimp measurement

During the earlier years in draw texturing, a relatively simple method was used in quality control for determining crimp value. This was developed by

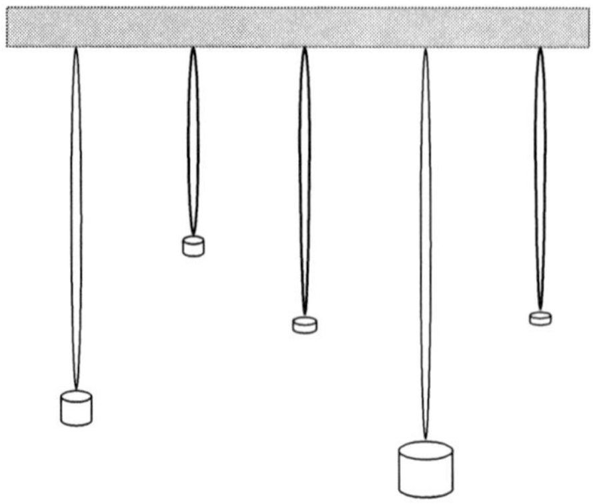

Load (cN/tex)	2	0.01	0.1	10 or 20	0.01
Duration of load (sec)	10	600	10	10	600
Length of hank	lg	lz	lf		lb

Crimp contraction% = ((lg-lz)/lg) x 100

Crimp modulus = ((lg-lf)/lg) x 100

Crimp stability =((lg-lb)/ (lg-lz) x100

10.6 Schematic diagram of load cycles for crimp contraction test.[4]

HATRA in the UK as a crimp rigidity test. Basically, the test method involved the application of a load to a yarn hank immersed in water and measuring the contraction of the hank on load removal. The test method is described in a recent article.[4] It is still used occasionally in the industry. However, the crimp contraction test procedure according to the German Standard DIN 53840 has more recently positioned itself as a global test method used in quality control laboratories.[5] By thermally treating a yarn hank of measured length, prior to applying pre-determined load times to the hank, differences in hank lengths as a result of the load application are automatically measured. From this, crimp contraction, crimp modulus and crimp stability are automatically calculated (Fig. 10.6). Crimp contraction is particularly important, as this is a measure of the 'bulking' of the yarn, affecting the fabric cover. It is largely influenced by twist level and heater temperature in the draw texturing process. The ability of the crimp to withstand load is represented by crimp stability: higher values, largely influenced by process tension, bring about increased crimp resilience to load.

10.3.3 Yarn intermingle measurement

For quality control of intermingling, statistical analysis of intermingle performance, including node distribution and maximum distance between nodes, are features in laboratory testing instruments. The equipment often

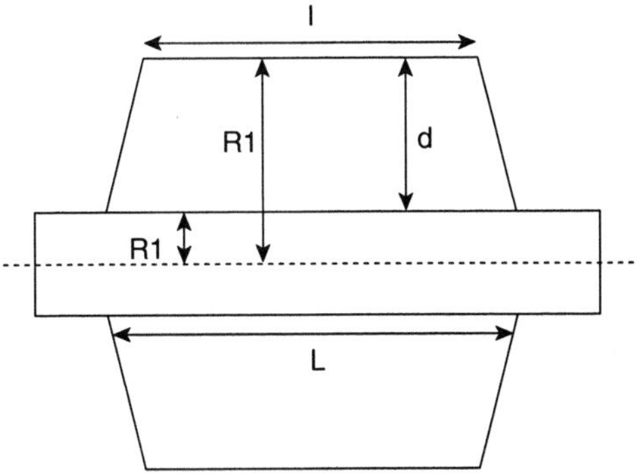

Weight of package = M1 grams
Weight of tube = M2 grams

R1 = package radius (cm)
R2 = tube outer radius (cm)

d= R1-R2 cm
Mass of yarn, M = M1-M2 grams

$$\text{Volume, } V = \pi d\left(\frac{(L+2l)d}{3} + (l+L)R2\right)\text{cm}^3$$

$$\text{Density} = \frac{M}{V}\ \text{g/cm}^3$$

10.7 Package density calculation.

has the facility to measure intermingle retention after subjecting the yarn to selected underfeeds (tension) within the test. Intermingle testing equipment that incorporates sensors for additional yarn testing requirements is also available on the market.[6] Here, wider applications include:

- Package unwinding performance, based on a statistical analysis of tension peaks during off-wind at pre-selected test speeds.
- Broken filament count.

10.3.4 Package density measurement

Package density is important in all process specification optimisations to ensure good package off-wind performance. For dye package production, however, where package densities are low to enable uniform dye penetration in the dye vessel, it is advisable to periodically check yarn package densities. Slight drift in raw material properties or texturing process conditions can significantly affect dye package densities. Test equipment is available on the market to measure package density.[7] Figure 10.7 shows how package density can be manually calculated.

10.4 Final product inspection and packing

Although automated package inspection lines are available in support of mass production in high labour cost countries, generally packages are still manually inspected and packed into cartons. Packing operators need to be well trained, with the grading criteria well defined and available at all inspection tables. They check all packages prior to packing for:

- Tail presence (for package transfer in downstream process).
- Package cleanliness and damage.
- Webbing.

Table 10.1 Summary of typical plant quality assurance functions

Control location	Test/control	Associated quality assurance areas								
		Raw material consistency indicator	Process specification definition	Position rectification	Production efficiency	Break rate	Quality efficiency	Product traceability	Records	SPC control charts
Machine	Monitoring:									
	shaft speeds		○							
	heater temperatures		○							
	Thread path alignment,			○					○	
	control and adjustments								○	
	Online tension monitoring	○	○	○	○	○	○		○	○
	Auto-doffing performance		○		○	○			○	○
At doff	Knit-dye test	○		○			○		○	○
	Label packages							○		
Laboratory (defined intervals)	Linear density (decitex)		○						○	○
	Tensile properties	○	○	○			○		○	○
	Crimp values	○	○	○			○		○	○
	Coning oil level		○				○		○	
	Intermingle		○	○			○		○	
	Broken filaments		○				○		○	
	Tight spots		○				○		○	
	Package density		○				○		○	○
Final inspection and packing	Package condition:									
	tail?			○			○		○	
	damage or oil stains?			○			○		○	
	broken filaments?			○			○		○	
	webbing?			○			○		○	
	intermingle?			○			○		○	
	Product identification on carton							○		
	Pack weights by grade				○				○	○

- Broken filaments, evident on the taper walls.
- Tube damage.
- Package size or weight.
- Intermingle nodes, where applicable, both on the yarn tail and outer layer.

Packages are graded according to the specified criteria and tolerances. Ideally, statistical records should be maintained of all faulty packages at the final inspection stage for traceability purposes and rectification of faulty processing positions. All cartons need to be clearly labelled with product reference, twist direction and weight, before palletising.

10.5 Statistical process control

It is good SPC practice to chart key test results by machine and product, together with process break rates, process efficiencies, knit-dye rejection levels and process tensions, for trend analyses to detect and take remedial action on:

- Progressive break rate increase, which may indicate heater cleaning need and determine cleaning frequencies.
- Drift in raw material quality, often highlighted by textured yarn elongation and process tension change.
- Drift in intermingle quality, which may be due to contaminated compressed air feed or contaminant build up in the air jets.
- Signs of gradual friction disc surface change (disc glazing, polyurethane disc wear) through process tension, tensile performance and crimp value drift.
- Package density inconsistency for dye package applications.
- Percentage package quality grading, based knit-dye test results, package size and, where online tension monitoring is fitted, tension faults.

A summary of typical test requirements and SPC chart needs for sound quality assurance practices is shown in Table 10.1. Performances should be reviewed over regular periods by plant management for product quality assurance, training requirements and remedial action needs.

10.6 References

1 Oerlikon-Barmag, Technical Brochure, *Solutions for the Draw Texturing Yarn Production*, Jun. (2008).

2 *The SmartTime Textile Processing Guide, Grey Scale for Assessing Change in Shade*, www.thesmarttime.com

3 A. Aygen, F. Avsar, *Lawson-Hemphill TYT-E*, Nov. 2009, www.lawsonhemphill.com

4 S. Mahish, R. Malik, S. Kumar, Effect of Process Parameters on Twist Textured Polyester Yarn, *Indian Textile Journal*, Apr. (2007).
5 *Texturmat ME*, Technical Literature, Textechno, www.textechno.com
6 *LabTEX Dynamic Yarn Testing Equipment*, Oerlikon-Fibrevision Textile Components.
7 www.lawsonhemphill.com
8 *Fibrevision Unitens*, Technical Literature, Oerlikon Textile Components, Apr. (2011).

11

Process costs in yarn texturing operations

Abstract: This chapter reviews process costs in yarn texturing operations. It discusses factors affecting costs and opportunities for yarn texturing outside low labour-cost countries.

Key words: false twist texturing, process costs.

11.1 Factors affecting costs

Manufacturing costs in draw texturing are expressed in cost/kg of yarn produced. The costs are very much influenced by:

- Machine output
  - machine efficiency
  - process speed
  - process break rate
  - quality downgrades.
- Price of raw materials.
- Labour costs.
- Energy consumption.
- Finance and depreciation costs.

In particular, finance and depreciation, labour and energy costs are dominant players in the total manufacturing cost of converting POY raw material to draw textured yarn. With little opportunity for significant speed increase in the draw texturing of mass produced commodity products, due to the fundamental constraints of the process, the main centres of the industry have relocated to low labour cost countries, such as China and India.[1] Moreover, the relocation has been closely followed by machine manufacturing, leading to machine availability at significantly lower prices than those of earlier European competitors on the local markets. Differences in energy costs are relatively small between key geographical locations[1] and raw material, labour costs and finance/depreciation costs are the main factors supporting relocation.

Of course, there is still a demand for 'quick response' to customer demands within the higher labour cost countries, such as for dye packages or spun dyed product for colour and for some technical end-uses where in-depth know-how is required. As referred to in Chapter 6, relatively high price machines that accommodate wider process flexibility and automatic package doffing have been introduced for such markets. These machines designs are aimed at lower energy consumption, lower labour requirements and high process flexibility. However, their manufacturing cost competiveness for mass produced commodity products against cheaper, non-automatic doffing machines in, for example, China, when achieving the same efficiencies and product quality, can be questioned. This is demonstrated in Fig. 11.1, which summarises an estimated yarn manufacturing comparison between automatic doff/high labour- and non-automatic doff/low labour costs in 2009. For the comparison, the assumptions shown in Table 11.1 were made:

These examples clearly show that labour and capital machine costs are significantly disadvantageous in Europe, despite labour saving from auto-doffing. Of course, there are other advantages of automatic package doffing in that exact equi-length yarn packages can be produced, which is particularly important in circular knitting and warp applications.

Table 11.1 Assumptions used for cost comparison, China vs Europe, Figure 11.1

Product	78dtex fine filament, nylon66	
Process speed (m/min)	650	
Machine efficiency (%)	95	
Break rate (bks/kg)	0.02	
No. of spindles	240	
	China	Europe
POY price(c/kg)	239	350
Energy price (c/Kw.hour)	7.5	9.5
Labour cost/hour (Euro)	0.83	15
Tube cost (cents)	12.6	17.1
Machine price (Euro)	240 000	750 000
Depreciation (years)	5	5
Calculated manufacturing cost (c/kg)	**273**	**439**
Calculated conversion cost (c/kg)	**26.1**	**46.1**

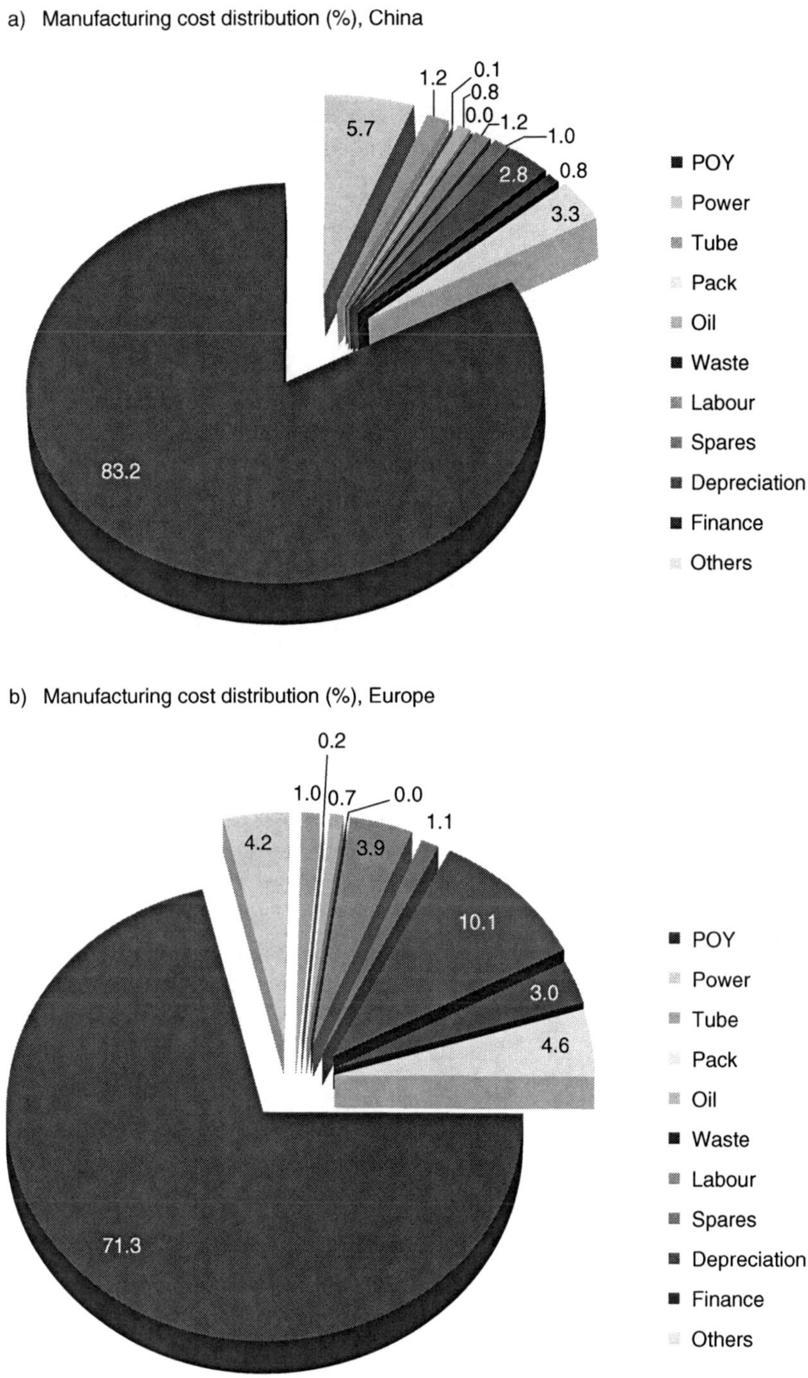

11.1 Examples of cost distribution (Eurocents) for manufacturing a nylon 66 78 decitex yarn.

11.2 Opportunities for yarn texturing outside low labour-cost countries

Where are the opportunities for draw texturing outside low labour cost countries? These can be considered as follows:

- Some branded garments require European yarn origin, so to some extent draw textured yarn prices are not confined to cheaper import prices.
- Where there are import duty constraints and significant transport cost advantages.
- Better product quality and quicker response to quality complaints.
- The opportunity for combined processes, such as draw texturing with air covering of elastanes, on the texturing machine.
- Use of older machines that are fully financed and depreciated, although the thread paths must be in good condition to support sensitive yarn production, such as microfilament yarns. This can, however, be considered a short-term benefit.
- Quick response to customer requirements for a wider range of products.

For a quick response to market needs, often involving relatively small quantities of products, process specification change is necessary, raw material yarn is usually required and often there is a need to change friction disc configuration and clean heaters. As a result, machine efficiencies, together with yarn conversion costs can be significantly affected. Figure 11.2 demonstrates the adverse effect of output efficiency loss on conversion cost. It also highlights the conversion cost advantage through the introduction of automatic package doffing and associated labour cost saving.

Comparative data is also shown for the production of low package density dye packages in Fig. 11.2, which doff at higher frequencies and are more labour intensive. Here, it can be seen that the introduction of automatic doffing brings about increased conversion cost saving. Indeed, when running a draw texturing plant, a focused attention in terms of cost saving must be given to:

- Manpower, automation and capital machine costs.
- Energy savings.
- Process optimisation and plant disciplines for highest efficiencies in both output and product quality.

An example of the effect of process break rate on yarn conversion cost is shown in Fig. 11.3. The proportion of product converted into Grade 1 quality affects the average sell price and hence profit margin for a specific process. When optimising a draw texturing process, quality downgrade and process break rate, both of which tend to increase with increase in process

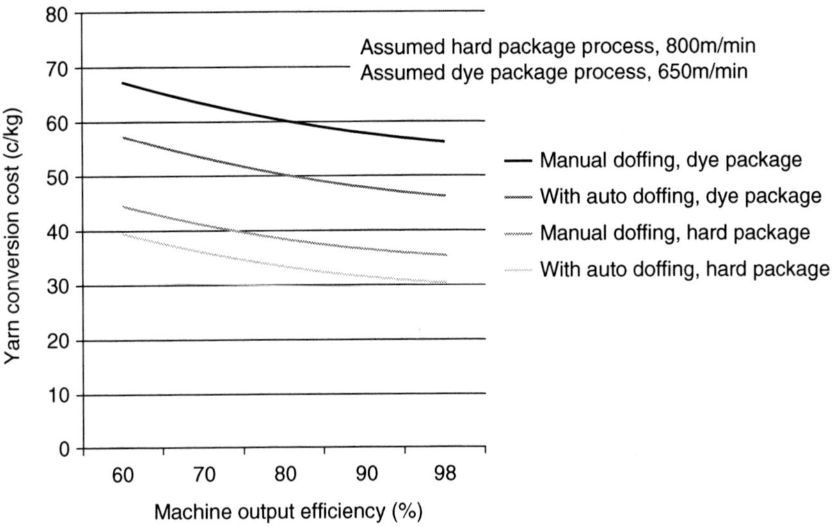

11.2 Effect of output efficiency on conversion cost (Eurocents) for a polyester 167dtex yarn.

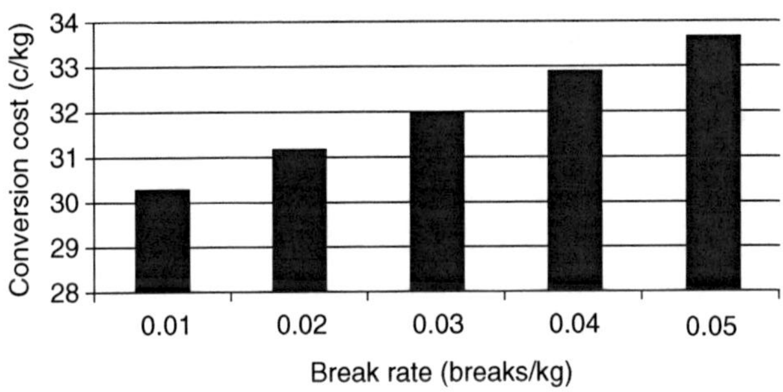

11.3 Effect of process break rate on conversion cost for a polyester 167dtex yarn (at 800 m/min, 98% machine efficiency and 92% POY tail transfer success).

speed, must be taken into account; it is not always cost effective to run processes at the uppermost speed that has been simply set on the basis of process instability limitations.

11.3 Reference

1 ITMF: Textured Yarn Production Costs, Chemical Fibers International, *Man-made Fiber Year Book* (2007).

12
End uses for draw textured yarns

Abstract: This chapter reviews end uses of draw textured yarns. It starts by discussing the beneficial properties of textured yarns before discussing applications of textured yarns made from nylon, polyester and polypropylene.

Key words: false twist texturing, nylon, polyester, polypropylene.

12.1 Property benefits of textured yarns

The draw texturing process imparts crimp to a continuous filament yarn, bringing about the characteristics of loftiness, stretch and recovery, good fabric cover, light weight, soft handle, dullness, durability, increased moisture absorption and low tendency to pilling when the yarn is in fabric form. The texturing process is flexible, enabling modifications to the yarn to be made so that yarns are adaptable to different end-use requirements. Yarn properties vary between the main thermoplastic draw textured yarns, polyester, polyamide and polypropylene (Table 12.1). As a result, each of these product categories tend to be more suited for particular fabric formation routes and end-uses.

In its early history from the 1940s to 1960s, polyamide yarns established a dominant position in textile end use due to their excellent recovery from deformation, which made them ideal for form-fitting garments where shape can be regained after stretching. Indeed, the texturing process was able to capitalise from these excellent stretch and recovery properties. Since then, polyamide textured yarns have developed and maintained positions in end uses where their key advantages could be utilised:

- Good abrasion resistance (luggage, socks).
- Good stretch and recovery (garment fit, hosiery).
- Softness and handle.

Table 12.1 Guideline properties of textured yarns

Property	Yarn type			Polypropylene
	Nylon 6	Nylon 66	Polyester, PET	
Specific gravity (g/cm^3)	1.14	1.14	1.38	0.91
Moisture regain at 65% RH and 20°C	2.5–4.5	3.5–4.5	0.2–0.5	0.04
Melting point (°C)	215–225	225–260	250–260	160–175
Dyeability	acid (and spun dyed)	acid	disperse (and spun dyed)	spun dyed
Glass transition temperature, Tg (°C)	40–87	46–90	80–85	−10
Tensile strength (cN/dtex)	4–6	4–6	3.5–6	2.5–6
Flammability limit oxygen index	20–21	20–21	20–22	19–20

As a result, polyamide yarns in particular have secured a position in end-uses such as in circular knitted hosiery, socks and, more recently, in seamless knitting, where stretch and recovery are important and the minimum amount of garment finishing is required.

PET polyester fibres on the other hand tend to have higher initial moduli than polyamide fibres and higher resistance to filament bending, leading to a crisper, less soft fabric handle.[1] PET polyester has, nevertheless, emerged as the more prevalent false twist textured yarn for a number of reasons:

- Cheaper and easier to produce (lower polymer costs, less complexity in POY spinning).
- Better light and wash fastness.
- Good crease resistance and dimensional stability in fabric.[1]

PET textured yarns tend to be used more in weaving applications where stretch and recovery of the fabric end use is not as important. Their scope, however, in apparel applications, has been broadened through development of finer filament yarns, degrees of lustre and filament cross-section variants, which have helped to improve fabric handle, minimise the 'synthetic look' and accommodated fashion changes.

12.2 Nylon

Nylon 6 and 66 textured yarns have good intrinsic properties:

- Good durability (abrasion and strength).
- Non-supporting of microbial growth.
- Good stretch and recovery.
- Low bending modulus for soft handle.
- Good fabric cover.

In particular, polyamide yarn stretch and recovery is superior to polyester: at 5% elongation they show 99–100% recovery against 75–80% for polyester.[2] This makes draw textured polyamide ideal for fabric applications such as hosiery, socks, active wear and work wear. Stretch and recovery is particularly important in leg wear for ladies' hosiery and polyamide has secured a position over many years in knitted fabrics for such end use. Moreover, polymer modifications, filament cross-section options and filament fineness and dullness variants have broadened fashion opportunities, creating better softness, drape, lustres, moisture management and colour fastness.

12.2.1 Ladies' hosiery

Circular knitting on four-feed, fine gauge knitting machines for ladies' hosiery is the largest application for low decitex polyamide draw textured yarn. Draw textured yarn can be supplied to the knitting process directly or via various intermediate process routes (Fig. 12.1). Because of their inherent good elastic recovery, polyamide yarns can be knitted as single yarns for hosiery. The tendency to finer filament yarns for improved softness and comfort, however, reduced the elastic recovery and today it is common to incorporate elastane in the knitted courses, either by combining it with the textured yarn in a covering process or by feeding it into the knitted structure by a plating or laid-in method (Fig. 12.2).

There are two intermediate process routes for covering elastanes with draw textured yarn (Fig. 12.3). Because elastic recovery is provided by the elastane, covering can also be applied to flat, drawn yarn for different appearance and sheer effect in the knitted garment. Conventional covering is a slow process whereby the pre-wound yarn is wrapped around the elastane at low speed. The elastane can be single-covered or double-covered at higher process cost to provide a softer fabric handle. Coning oil levels applied in the texturing process tend to be between 0.5% and 1.0% for conventional covering application; too high an oil level creates oil splash, which can cause slippage of the package drive mechanisms in covering. Air covering is typically carried out directly from the textured yarn package at

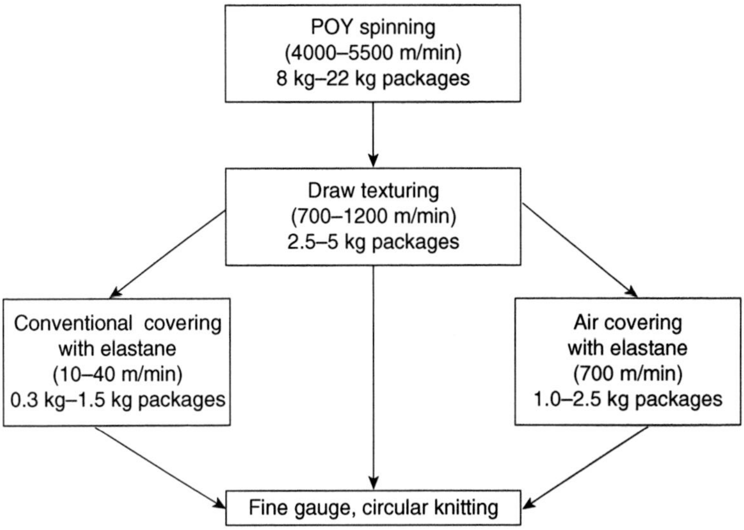

12.1 Polyamide draw textured yarn routes to hosiery knitting.

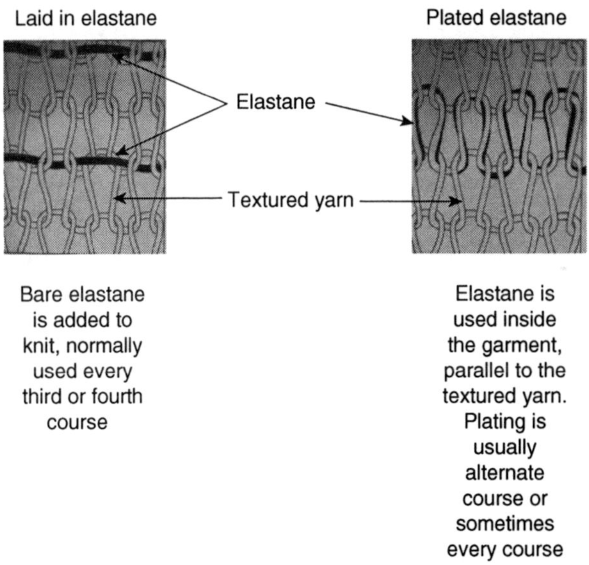

12.2 Concept of laid in and plated elastane in knitting.

speeds around 700 m/min; it utilises air jets to intermingle the filaments of the yarn with the elastane core (Fig. 12.4). Air-covered elastanes tend to be harsher in handle than their conventionally covered counterparts. Figure 12.5 shows an example of an air-covered textured yarn in the knitted hose. In recent years, machine manufacturers have introduced the facility to

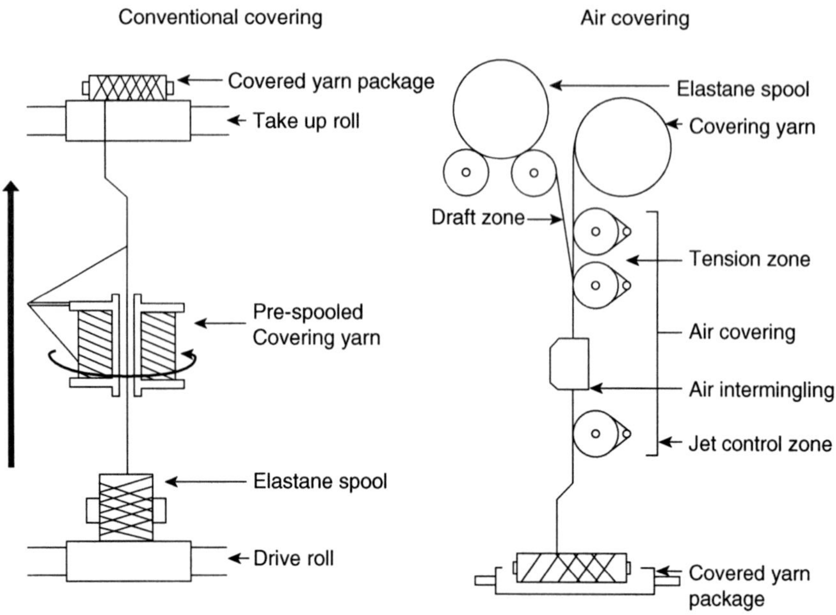

12.3 Elastane covering processes for draw textured yarns.

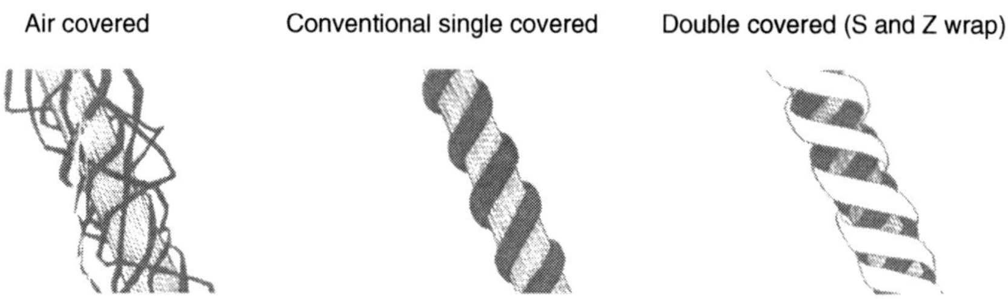

12.4 Concept of conventional and air-covering elastanes.

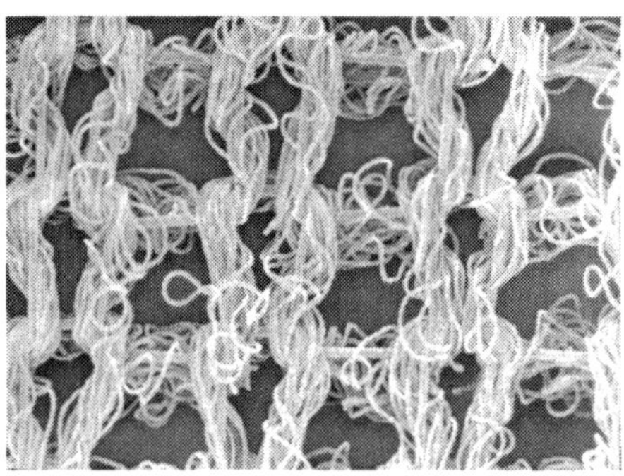

12.5 Air-covered draw textured nylon 66 20 dtex f22 with 11 dtex elastane.

12.6 BTSR (Italy) elastane tension control device.[6]

integrate the air covering process into the draw texturing operation, thus providing a single-stage process (See Chapter 6).

Nylon 6 is cheaper, easier to dye and consumes less energy than nylon 66 in textured and dyed yarn manufacturing. It has, however, inferior stretch and recovery, but with the use of elastanes, helps to overcome this disadvantage in hosiery end use. Hosiery knitting speeds have increased significantly over the last two decades, with speeds exceeding 1000 revolutions per minute, thus imposing higher quality demands on false twist textured yarns. In particular:

- Coning oil type and level have become more critical for low yarn friction consistency.
- Package build is more important to maintain low tension peaking from filament snagging and avoid stoppage in knitting.
- The increase in use of microfilament yarns increases tendency to tension peaking at package off-wind and needle breakage due to filament flaring and trapping in the needles. This can be overcome by the use of a low intermingle level in texturing.
- When plating, elastanes, can move to the front of the knitted stitch due to tension variation, causing a flash appearance. To a large extent, this has been resolved by the use of online tension control damping on the elastane feed (Fig. 12.6), but there remains an emphasis on the need to optimise the texturing conditions for consistency in feeder yarn tensions.

12.2.2 Socks

The manufacture of half hose (socks) has traditionally been a major consumer of false twist textured polyamide yarns. Usually, 20% by weight of a cotton or wool sock comprises textured polyamide yarn. Textured polyamide yarn provides sufficient stretch for multi-size products and enhances durability. There is some use of elastanes in socks for broadening fit sizes, but this remains a relatively small proportion, due to a large extent to added manufacturing costs; for the elastane to be useable, it has to be covered with polyamide via a conventional or air covering.

12.2.3 Circular knitting

The Santoni seamless technology has been a significant advance in circular knitting and has proved to be ideally suited for polyamide yarns because of available variants, such as filament fineness, lustre and filament shape, together with their high stretch and recovery. The technology is based on knitting tubular fabrics without seams and can apply different knit stitches, incorporating pre-shaped structures, avoiding costs, such as manpower, waste, fabric stocks and energy, which are associated with the alternative 'cut and sew' manufacturing method. Seamless technology has expanded at a rapid rate, with end-uses in lingerie, swimwear, sportswear, underwear and general fashion garments. Between 2001 and 2003, seamless underwear products increased from 9% to 18% of global production.[3] Today, this figure is believed to well exceed 35% of global production.

Medium decitex polyamide yarns, 44 dtex, 78 dtex and two-fold 78 dtex, are also traditionally used in coarse gauge circular knitting for apparel end-use. The machines are typically equipped with 72 or 96 feeds for plain, pattern effects and with plated elastane capability. Usually, the machines are fitted with yarn feed storage devices for maintaining steady feeder tensions. Regarding false twist textured yarn requirements:

- Packages should be equi-length for off-wind tension consistency across the feeds.
- Package build should be optimised for low tension peaking at off-wind (wind angle, density and edge disturbance).
- Highest possible crimp stability is required on finer filament yarns to avoid influence from any tension variation.
- Light intermingling is preferred, particularly on finer filament yarns, to reduce tendency to filament snagging in package off-wind and at the needles.

12.2.4 Yarn colouration

Increasing labour costs and increased yarn conversion costs have necessitated West Europe to explore means of cutting costs in their process routes over the last several years. This is particularly true where low-density dye packages are concerned, as doff times tend to be short and more labour intensive due to relatively smaller package sizes. Moreover, polyamide textured yarns have inherently high shrinkage when exposed to dye temperatures in the order of 95–100°C, and to maintain package stability (colour consistency and bulk), dye packages are produced in the draw texturing process at extremely low packing densities. In addition, air intermingle has tended to replace twisting in Europe, as a further process stage elimination, although this is to the detriment of fabric handle.

For the polyamide false twist process, machines tend to be single-heater versions, purposely designed to capitalise on the high bulk/elastic recovery properties of polyamide ecru yarns. Therefore, unless the processor also textures polyester yarns, there is limited access to set heater (twin-heater) machines. For draw textured polyamide yarn dye packages, however, twin-heater machines can be advantageous, offering more flexibility to:

- Pre-stabilise the yarn, reducing shrinkage in the package dyeing operation, leading to more uniform dye uptake within the package.
- Realise higher package densities, potentially eliminating the need for a re-wind process after dyeing, i.e. direct use of dyed yarn on plastic dye tubes after dyeing.
- Achieve more dye uniformity on finer filament yarns within the packages due to reduced yarn shrinkage.

Process routes for false twist texturing and package dyeing are summarised in Fig. 12.7. Nylon 6 can also be solution dyed in POY spinning. Older traditional process routes for dye package production still apply (Fig. 12.8). These overall processes are labour intensive. In terms of yarn properties, however, high elastic recovery can be achieved via the low-density muff dye route.

12.2.5 Other end uses

Polyamide false twist textured yarns are also used in warp knitting, flat bed knitting, weaving, narrow weaving and in technical weaving applications, such as textile members for drive belts.

12.3 Polyester

False twist textured yarns manufactured from polyethylene terephthalate (PET) are used extensively in woven fabrics for apparel and home furnishings. Whilst synthetic clothing can be perceived to have a lower natural feel compared with fabrics woven from natural fibres, polyester fabrics can provide good wrinkle resistance, durability and high colour retention.

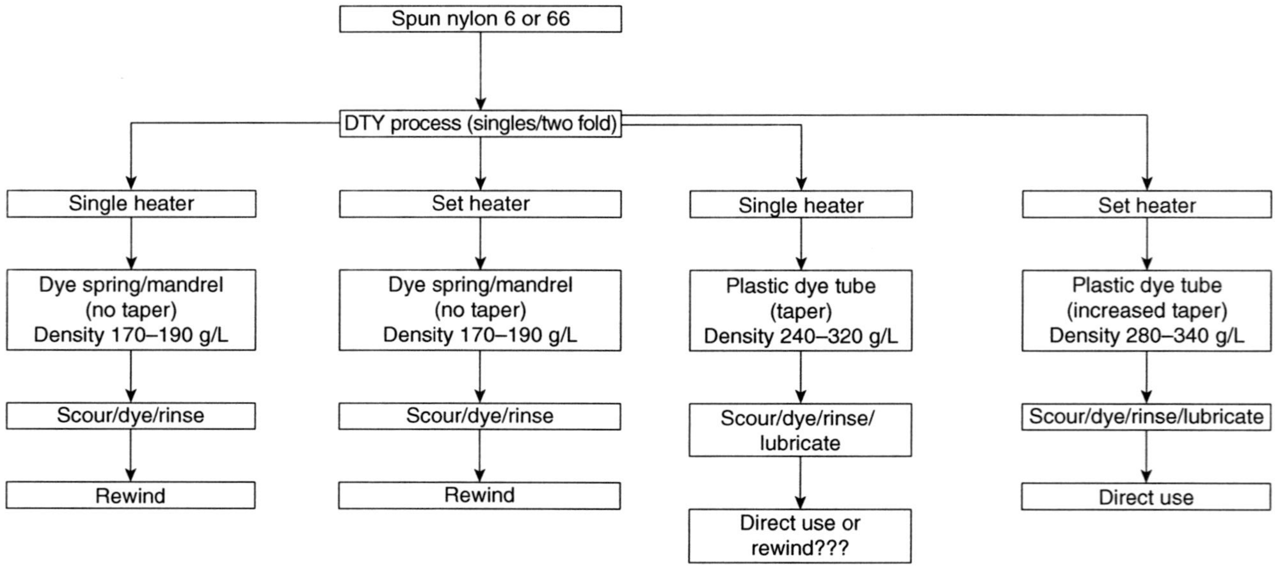

12.7 Major routes for polyamide dye package manufacturing.

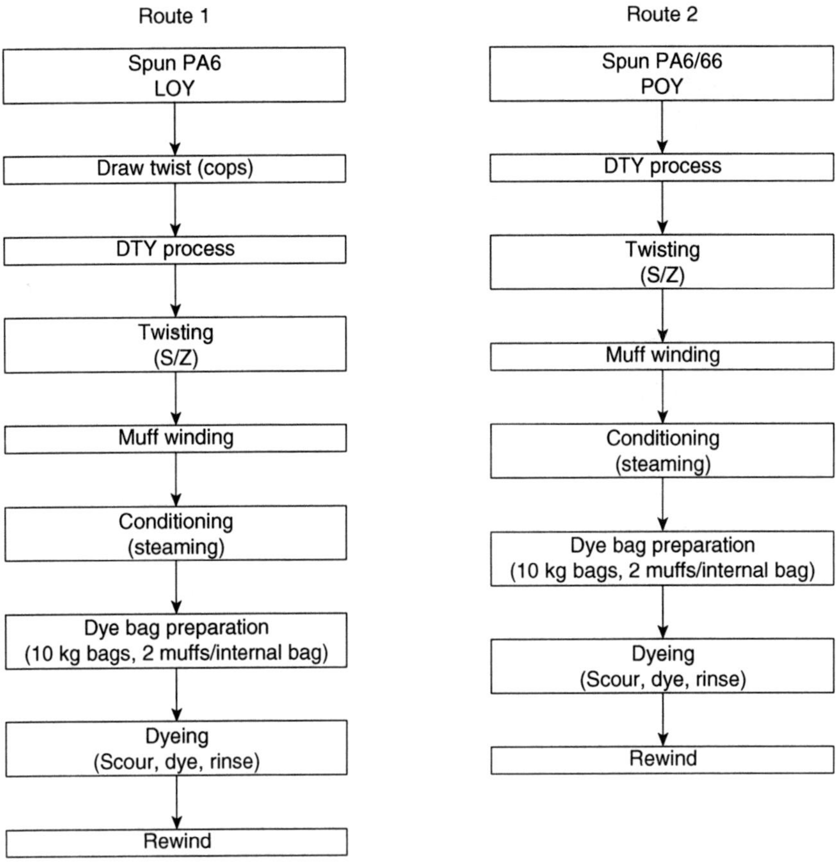

12.8 Older, traditional process routes for polyamide draw textured dyed yarns.

12.3.1 Weaving

Weaving is the largest application for false twist textured products. The weaving process interlocks two sets of threads, one running lengthwise along the fabric (warp) and the other running across the fabric (weft) (Fig. 12.9). Each individual warp thread is called an end and each individual weft thread is called a pick. Textured yarns can be used as both warp and weft yarns. The selection of the yarn interlocking pattern determines the design or weave of the fabric. This, in combination with the number of ends and picks per centimetre and yarn linear densities, determine the structure of the fabric. In weaving, the weft yarn is carried across the selected openings of the warp yarn (shedding) via shuttle, rapier, projectile gripper, air jet or water jet.

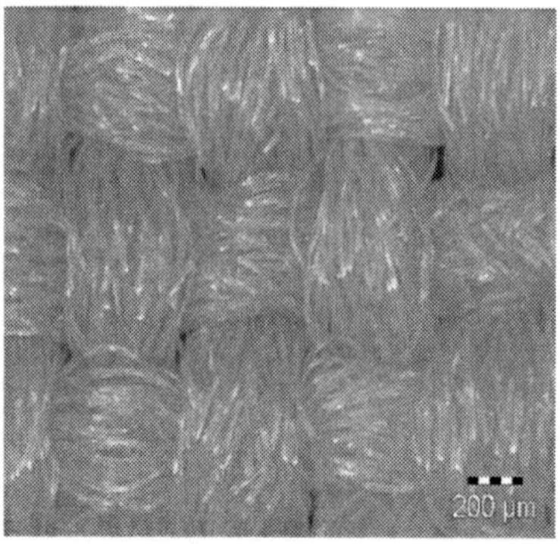

12.9 Simple 1/1 twill, comprising false twist textured polyester yarn.

Over the last 30 years, a 300% increase in weaving performance has been realised. This advance is attributable to weft insertion modifications in single-phase weaving. As speeds have increased, load bearing limits on the yarn due to the high acceleration at weft insertion have similarly increased. As a result, higher product quality demands are placed on the yarns in terms of:

- Tensile properties.
- Broken filament levels.
- Frictional properties.
- Package off-wind performance.

Package off-wind is assisted by the use of weft accumulators, which periodically draw and accumulate the yarn from the packages and insert into the weft at more uniform tension. In the 1990s, high weft insertion rates in weaving were possible with speeds of up to:

- Rapier: 1000 m/min.
- Air jet: 1500 m/min.
- Water jet: 1750 m/min.

Today, the weaving industry has developed multi-phase weft insertion at speeds of up to 6000 m/min. Speeds tend to be set according to the off-wind limitations of the packages, which for four simultaneous weft insertions equates to a 1500 m/min for a total 6000 m/min insertion rate.

Warping is carried out by the rewinding of yarn packages onto multi-thread holders, known as section warp beams. In the process:

- Equi-length packages are creeled onto a section warper that can have speed capabilities up to around 750 m/min.
- Back beams for weaving are wound from the section warp beams for sizing or direct use as a non-sized product on weaving machines.

The purpose of sizing is to bind the filament bundle, so that protruding filaments or loops are not problematic in the weaving operation. Sizing, however, brings about the following problems:

- It is an extra process, adding cost to the process chain.
- The size (essentially starch) has to be removed after weaving through a fabric scour. Waste from the scouring operation has to satisfy environmental legislation and is coming under ever increasing scrutiny.

Where intermingling nodes are not visibly detrimental to the woven fabric appearance, intermingling air jets are used in draw texturing to produce warp yarns for a non-sizing process route. This method of binding the filaments is commonly used today (see Chapter 7, Air jet intermingling). For this application, it is important that the yarn comprises hard nodes, with low occurrence of non-interlaced yarn ends. Twisted false twist textured yarns can similarly be used, but the intermediate process of twisting adds significantly to the yarn manufacturing costs. For warp application too, false twist textured polyester yarns tend to be set using the secondary heater in texturing. High crimp values are not so important here, since woven fabrics do not demand high elasticity, and set yarns prove extremely beneficial in warping in that twist liveliness due to residual torque can be eliminated.

Advances in polyester yarn developments have further improved its position in woven fabrics through:

- More effective use of microfilament yarns for improved comfort and fabric handle.
- Modified lustres and filament cross-sections for reducing brightness and 'the synthetic look'.

Microfilament yarns demand much emphasis on process optimisation in false twist texturing to satisfy the high speed demands in warping and weft insertion. The yarns must be broken filament free to avoid slub formation and loom stoppage, package-build specifications must be carefully optimised for trouble free package off-wind and crimp volume and tendency to filament flaring should be minimised through use of air intermingling jets.

12.3.2 Warp and Raschel knitting

In warp knitting, the needles produce parallel rows of loops simultaneously; these are interlocked in a zigzag pattern (Fig. 12.10). Warp knitted fabrics are produced in a sheet form using one or more sets of warp yarns. The yarns are fed from warp beams to the needle row that extends across the machine width.

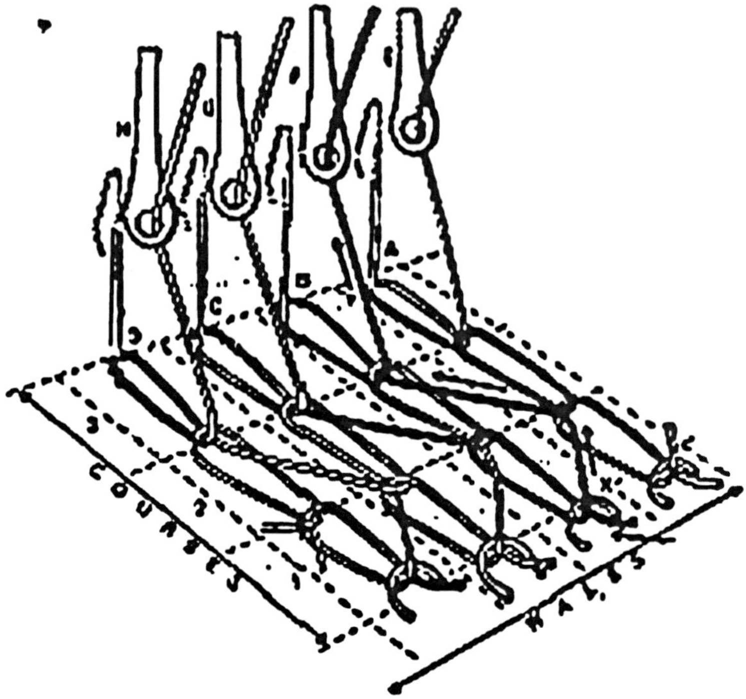

12.10 Warp knit concept.

Basically, the stitch is formed via the swinging movement of the yarn guide bars and the up and down movement of the needles. For warp knitting, textured yarn packages should be equi-length and typically the warp comprises 1170 to 1500 packages for 28 gauge to 40 gauge machines (gauge = number of needles per inch) respectively.

The main knitting methods applied are Tricot and Raschel. The Tricot machine has 2–5 bar combinations, whereas the Raschel machine ranges for 3 bar to 78 bar combinations for better engineering of yarn type and varying degrees of directional fabric stretch. Tricot fabrics tend to be soft, wrinkle resistant and have good drape. They are used in lingerie, sleepwear, shirts and dresses, etc. Raschel fabrics cover a wider range of end uses including lace, swimwear, sportswear and vegetable bags. Similar requirements are imposed on yarns for warp knitting as in warping for weaving; basically, the yarn-associated problems occurring in the pre-knitting operation, warping. Multi-ends are beamed at speeds of up to the order of 800 m/min, creel management being of particular importance.

For warp knitting, textured yarns are usually:

- Set yarns, applying a secondary heater in the textured yarn process, to reduce residual torque.

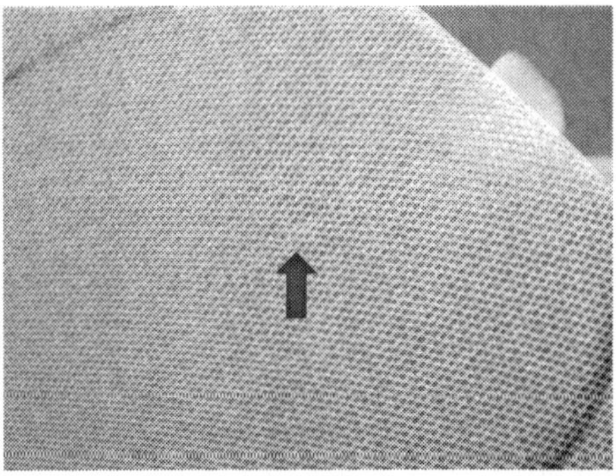

12.11 Slub in warp-knit fabric due to broken filament formed in warping.

- Intermingled, to avoid tendency to slub formation (Fig. 12.11). Too high an intermingle intensity, however, can create pinhole/pattern effects in open-knit structures.

In addition:

- The textured yarn must have no broken filaments inherent from the texturing operation, and must display good and consistent tensile performance.
- Coning oils in texturing should be selected for low friction.
- For two-fold yarns, intermingling should be such that the yarns do not split between nodes, as this tends to create small snarls, which are carried into the knitting process.

12.3.3 Circular knitting

Polyester false twist textured yarns are also used in circular knitting, where the addition of elastane enhances the fabric stretch properties. Typical end uses are for sportswear and leisurewear. Circular knitting can be quite problematic in terms of fabric appearance, where slight crimp variation, yarn frictional properties, intermingle consistency and crimp stability can adversely affect both dye shade and fabric cover. Attaining adequate crimp stability on microfilament yarns to minimise influence from any tension variation in the knitting process requires particular attention. Some typical faults experienced in circular knitting and their sources are listed in Table 12.2.

Polyester has good ultra violet light fastness properties and is used in automotive textiles. Automotive textiles for door coverings and seat fabrics are commonly manufactured from polyester false twist textured yarns on circular knit machines. Polyester false twist textured yarns are also used

Table 12.2 Examples of faults experienced in coarse gauge circular knitting with false twist textured yarns

Fault description	Cause	Source/Rectification (Examples)
Line barre, single courses	Yarn tension (prominent in transmitted light)	Knitting feed tension variation/friction Poor package build optimisation Inadequate yarn lubricant consistency Oligomer/ friction effects (dye packages)
	Dye uptake variability (prominent in reflected light)	Process tension variation in texturing Thermal history variation in texturing POY quality variation
	Bulk variation (evident in both transmitted and reflected light)	Yarn feed slippage in texturing POY quality variation Disc glazing Process too near instability speed Inadequately optimised disc combination Intermingle inconsistency Poor crimp stability
	Random stripes (tension transients)	Short-term tension transients, usually POY related
	Periodic stripes (surging)	Inadequate process optimisation (surging) Input feed slippage Variation in POY draw force
Block barre, multi-adjacent courses	Stitch size variation (tension, stitch length) evident in both transmitted and reflected light	Normally knitting cylinder alignment related Reset dial and cylinder levelness in running
	Fabric draw down variation	Stop/start variation in knitting machine Variation in draw down of fabric, but on modern machines is controlled by an automatic friction roll system for constant tension

in warp knit for automotive applications, such as for car head liners. Air jet textured yarns tend to be used in woven fabrics for automotive applications, such as for car seats, where their good abrasion properties are advantageous.

Because of quick response needs for colour in the automotive chain, false twist textured yarns have historically been of dye pack origin for circular knitting. Today, however, despite minimum batch size limitations (colour quantity restriction), there is an increase in the use of coloured textured yarn from the solution dyed route in spinning. This process route is more cost effective, eliminating the separate package dyeing process.

The whole supply chain process for circular knitting for automotive end-use is extremely quality orientated. In knitting:

- The draw textured face yarn is fed normally on 72 or 96 feeds to the circular knitting head via yarn feed accumulators. Off-wind speeds are typically up to 600 m/min, consistency in yarn tension and friction being paramount if fabric appearance is to be trouble free.
- In knitting, the textured face yarns are knitted together with a backing yarn, which is commonly spun dyed false twist textured polyester.
- Low profile loops are created with the face yarn (Fig. 12.12).

The fabric is then slit and wound onto preparatory rolls for finishing. The face loops are cropped under fabric tension, usually twice, before having laminated backing material applied at temperatures around 135°C (Fig. 12.13).

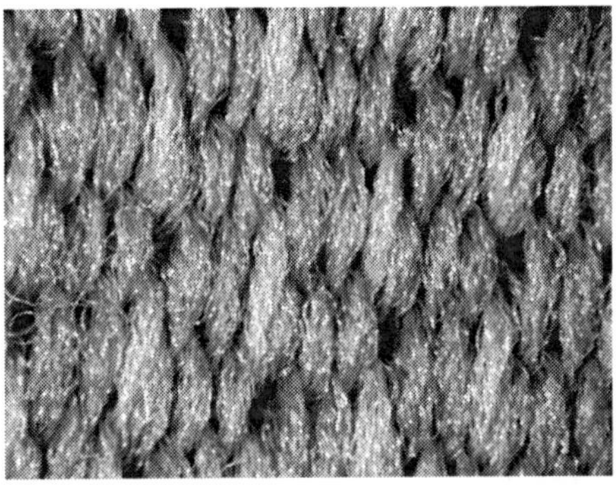

12.12 Knitted loop structure for automotive pile fabric.

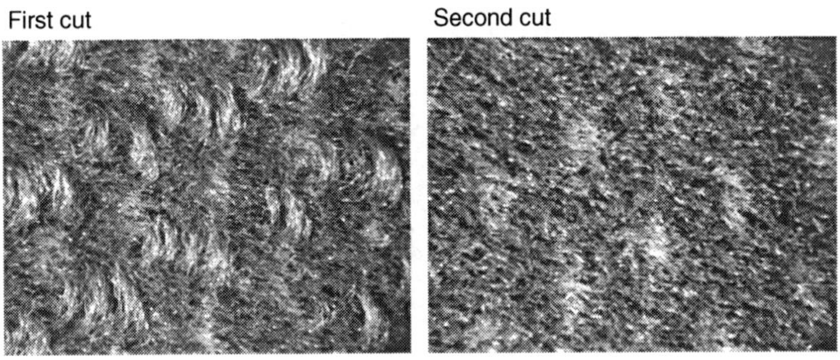

12.13 Cropped automotive circular knit velour.

Velour fabrics are used for particularly critical applications and impose high quality demands on the false twist textured yarns. For example:

- Residual torque variation in the yarn can affect the pile light reflection giving the impression of dye shade variation.
- Dye shade difference due to liquor penetration problems at the traverse reversal area on dye packages are experienced. Dye package build in texturing has to be optimised for soft edges.
- Where dyed packages are directly used and not subject to a re-wind process, good off-wind characteristics are required to minimise tension peaking. Lubricant type and level consistency within the dyed package is important for low yarn friction and friction consistency. Lubricant is applied in the dye vessel after dyeing.
- Where air intermingling is applied in the draw texturing process to enhance package off-wind performance, too high a level of intermingling can create a pinhole appearance in the fabric face.

12.3.4 Colouration

For the dyeing of polyester dye packages, oligomer re-deposition on the yarn in the dye bath tends to create high and variable yarn-to-ceramic friction, which causes tension variation in the knitting process in automotive circular knitting applications. This is particularly critical, as it can lead to stripes (optical) in the pile fabric. Oligomer deposition can be particularly problematic (Fig. 12.14). Oligomer is a low molecular weight polymer, containing only a few monomer units. It can be reduced by:

- High temperature drain of the dye liquor after the dye cycle.
- Alkali dyeing (a relatively new process), although there are some limits with respect to colours attainable.

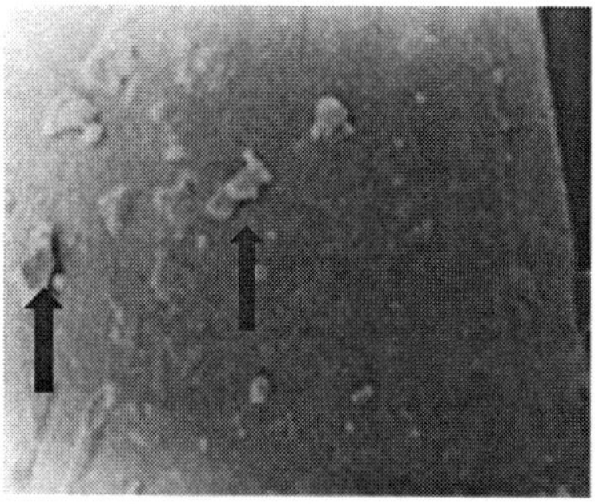

12.14 Oligomer on surface of a polyester filament (5500 × multiplication).

- Rewinding dye packages, where a high quantity of the oligomer is taken off by friction on the contact surfaces in the rewind process. This adds yarn manufacturing cost. Re-oiling the packages in the rewind process can also provide better protection against oligomer being deposited on guide surfaces in knitting.

12.3.5 Other end uses

False twist textured polyester yarn is used in sewing threads, where high demands are placed on achieving uppermost yarn tenacity and intermingle quality. The yarns also have applications in flat bed knitting for apparel, and narrow woven fabrics used for apparel, braiding, ribbons, furnishings and technical applications.

12.4 Polypropylene

Polypropylene has good resistance to staining and chemical attack. It has a low specific gravity compared to polyester and nylon (Table 12.1) and, as a result, the same fabric in polyester would be around 30% heavier. Being solution dyed in spinning, it has excellent colour fastness and it is also hydrophobic, displaying good moisture wicking properties. The additions of master batches for colouration in spinning has influence on the properties of the yarn, such as tensile performance and friction.[4,5] The positive characteristics of polypropylene false twist textured yarns are:

- Light weight due to low specific gravity.
- Resistance to deterioration from chemicals, mildew, insects and perspiration.
- Good abrasion resistance.
- Low moisture absorption.
- Stain and soil resistance.
- Low static build-up.
- Good washability, quick drying.

The disadvantages of polypropylene are:

- Low melt temperature, which limits ironing.
- Cannot be dyed by conventional methods and is largely limited to solution dyeing in spinning.
- Tends to creep under tension due to its low glass transition temperature.
- Poor UV and thermal stability requiring expensive UV stabilisers and antioxidants to overcome this problem.

Polypropylene largely finds its end-use in nonwoven textile applications, particularly for hygiene products. In the false twist textured form, it is used for sportswear, socks and outdoor wear, where its moisture wicking and thermal properties can be used to full advantage.

12.5 References

1 J. G. Cook, *Handbook of Textile Fibres, 2. Man-made Fibres*, Merrow Publishing Co. Ltd, 1968.
2 S. S. Mahish, S. K. Ladda, PTT–Fiber of the Future, *Cemical Fibers International*, 54, Oct. (2004).
3 *Santoni–the seamless story*, www.santoni.com
4 R. Raghavendra, R. Hedge, A. Dahiya, M. G. Kamath, www.engr.ukt.edu/mse/pages/textiles/olefin%20 fibers.htm.
5 M. Vanneste, S De Decker *et al.*, Effect of color pigments on PP FDY yarn properties and weaving behavior, *Man-Made Fiber Year Book*, 2006, Chemical Fibers International.
6 www.BTSR.com

13
The future of false twist texturing

Abstract: This chapter reviews future trends in false twist texturing. It discusses global production trends, machine development, process integration and raw materials.

Key words: false twist texturing, global production trends.

13.1 Global production trends

Polyester textured yarns represent a significantly higher worldwide consumption than polyamide textured yarns. With their low labour costs, China and to a lesser degree India, have emerged as the largest manufacturers of PET textured yarns. In 2005, for example, it is believed that China accounted for about 52% of the global production of PET synthetic fibre, and over the last two decades the manufacturing of false twist textured yarns has markedly increased within that country. Figure 13.1 highlights this rapid growth.[1] China and India accounted for around 90% of the false twist texturing spindles supplied to the polyester yarn industry in 2004 and 2005 (Table 13.1).

Although Western Europe has maintained a position in the production of polyamide false twist textured yarns, which has been influenced by the strength of the ladies' hosiery industry in Italy, growth is also evident in China (Fig. 13.2).[1] Similarly, in recent years, it is believed that sales of single-heater draw texturing machines for polyamide yarns have been concentrated in Asia. China, accounted for approximately 62% of the total worldwide spindle sales in 2004/2005 for this end-use (Table 13.2). Since then, investments in single-heater draw texturing machines for high stretch polyamide yarn applications have significantly reduced in Asia, with an estimated 2000–3000 spindles being sold in China for 2007/2008.[2]

China has been predicted to currently produce approximately 150 000 tonnes per annum of polyamide textured yarns in 2010, which equates to an order of 20% of forecast global production of textured yarn in 2010. The

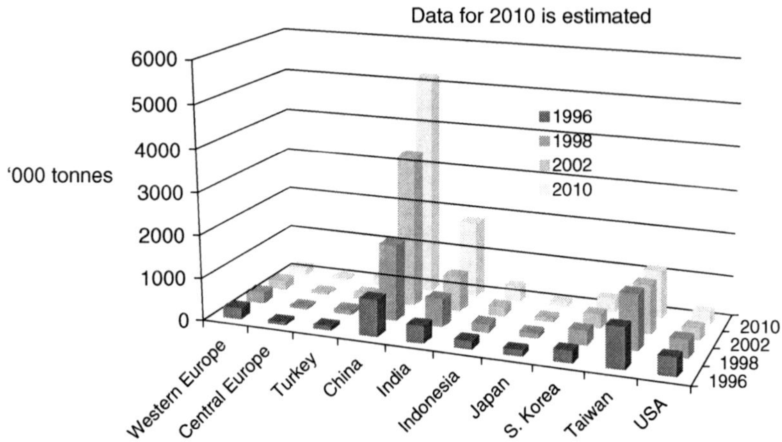

13.1 False twist textured polyester yarns; global manufacturing trends (source, CIRFS, 2004).

Table 13.1 Supply of false twist texturing spindles for polyester application

	2004	2005	% share (2 years)
Spindles (total)	392400	308330	
China	340800	259956	85.7
India	18336	13716	4.6

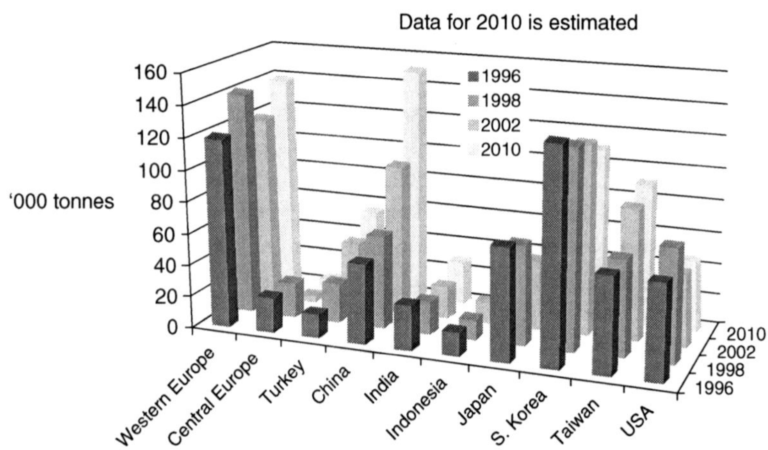

13.2 False twist textured polyamide yarns; global manufacturing trends (source, CIRFS, 2004).

Table 13.2 Supply of false twist texturing spindles for polyamide application

	2004	2005	% share (2 years)
Spindles (total)	13200	6872	
China	9048	3384	61.9
W. Europe	0	1200	17.5

other major producer is Western Europe, which is believed to account for approximately 15% of global activity. In the Far East, including China, nylon 6 historically has been the major polyamide consumed in texturing because it is cheaper to produce and easier to dye. Moreover, spun dyed colours are a possibility with nylon 6: with nylon 66, current technology limits spun dyed colours to black and white. Over the last two years, however, there is some evidence of small growth in the production of nylon 66 in China, thought to be mainly for export in seamless, circular knit and hosiery applications.

13.2 Machine development

False twist textured yarns bring about desirable properties in fabrics, such as handle, softness, elasticity, cover and moisture- and air-permeability. Their mechanical properties have also proved desirable in certain technical end-uses. The process has established itself over the years as an important integral part of the textile manufacturing chain and these desirable yarn characteristics are unlikely to be realised by alternative methods in the foreseeable future.

The false twist texturing process will, of course, remain a subject of development, both in machine design, product innovation and for accommodating raw material developments or variants. However, after a history of much research and development, it is now recognised that a significant step increase in speed cannot be realised. This is largely due to limitations imposed by the onset of process instability. In certain applications too, high-speed air intermingling also limits the uppermost speed attainable in the process. Of course, advances have been made over the years in the heat–cool lengths and angles of wrap in the texturing zones to raise instability speed thresholds, but the underlying limitation that restricts significant speed advance remains within the false twist concept itself.

Because of these constraints, false twist texturing machine development will most likely continue to focus on the design of more cost-effective processes based on energy, maintenance and labour saving, and with wider scope for process flexibility for processing a broader range of yarn variants and types. This view is typified by recent developments of Oerlikon-Barmag with their eFK machine,[3] which amongst other features, incorporates godet yarn feed systems for low energy consumption, higher operator efficiency, reduced maintenance costs and lower noise emission. Machines are likely to increasingly evolve in modular build form for accommodating both high volume output and small, flexible output needs. Similarly, it is likely that they will be designed for the add-on capability of multi-fold, double density and yarn combination needs.

It is also safe to assume that machine manufacturing will remain in the lower cost countries and in close proximity to the industry so that capital investment costs for machine investments can be contained. Progressively, labour costs are likely to rise in these countries, leading to increased opportunities for the introduction of automatic doffing facilities. The rapid relocation of the draw textured yarn manufacturing industry to China in recent years has indeed brought about a similar relocation of machine and component manufacturing. Currently, these machines are largely copies of machines that evolved from the high development activities within Europe and Japan in the 1980s. Short-term advance in false twist texturing machine developments is unlikely, but after a period of time, with progressive increase in process know how, it can be assumed that further innovation could well materialise from within China and India.

13.3 Process integration

Today, mechanical speed capabilities of false twist texturing machines are up to 1500 m/min and it is certainly not beyond technology to increase this mechanical capability, with even consideration given to integrate the texturing process into the melt spinning line, thus reducing process stages and potential costs. However, the constraints of the false twisting concept still remain and loss in polymer throughput can be considered cost prohibitive. Long term, there is always a possibility that self-crimping side-by-side bicomponent filament yarns,[4] which comprise two polymers in spinning, could be a small threat to false twist texturing. However, it is unlikely that the crimp characteristics and handle of the false twist textured yarns can be fully realised by this alternative route.

Over the years, there have been attempts to integrate some downstream processes with that of false twist texturing. For example, Karl Mayer introduced a novel machine to the market for drawing, texturing and warping POY in a single stage process.[5,6] Further integration has successfully centred on the air covering of textured yarns with elastanes and false twist texturing machines are widely available with air covering facilities today.[7,8] The extent of further development here depends on the willingness of the market to more widely substitute conventional covered yarns with the cheaper but harsher-handle air-covered products.

13.4 Raw material yarns

The extent at which the false twist process itself can be significantly changed to meet fabric property or specific end-use requirements is limited: basically, the processor is restricted to changing the crimp characteristics of a yarn or its tensile properties. With the exception of effect yarns, brought about by yarn combinations, changes to appearance, handle and characteristics largely stem from the raw material yarn. Here, for example, through polymer type and filament cross-section, lustre can be determined, and through filament numbers and fineness, softness and fabric handle can be influenced. Over the last 15 years, new yarn types with various functionalities have emerged from fibre producers. These variants stem from additions to polymers before spinning. They include, for example, anti-bacterial, anti-odour, controlled release of skin emollients and ultra violet protection properties.

As alternative to polyethylene terephthalate, polybutylene terephthalate (PBT) has historically stimulated interest in textile apparel applications because of its higher elasticity. However, apart from its use in special effect yarns brought about by crimp, elasticity and dyeing, it has not really made a significant impact. It remains commonly used in mouldings because of its good mechanical strength, low melt temperature (215–225°C), low electrical conductivity and low moisture absorption.

Of more interest, perhaps, is polytrimethylene terephthalate (PTT), which is another aromatic polyester, made by the polycondensation of 1,3-propandiol and terephthalic acid (Fig. 13.3). PTT has good stretch and recovery, softness and can be dyed at low temperatures. It is claimed to have the best properties of polyester and nylon in textile applications.[9] Whether this polymer type will secure a position in false twist texturing remains to be seen.

More recently, polylactic acid (PLA) has stimulated much interest in the industry. PLA is made from the starting material, lactic acid, which is formed by fermenting natural sugars from agricultural crops, such as corn or sugar beet.[10] PET and PTT are aromatic polyesters with benzene rings in their repeat units, whereas PLA is an aliphatic polyester (Fig. 13.4). PLA is suitable for textile apparel applications because:

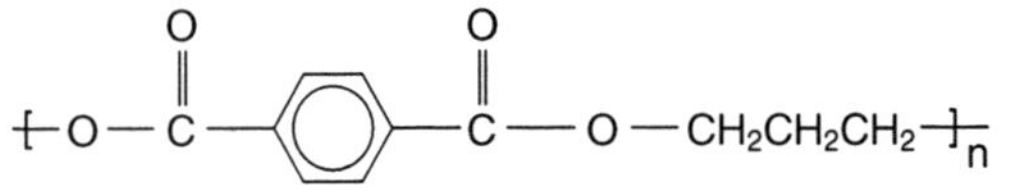

13.3 Structure of polytrimethylene terephthalate.

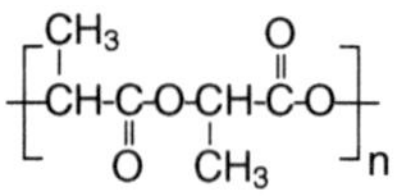

13.4 Structure of polylactic acid (PLA).

- It has low moisture absorption and good moisture wicking properties.
- High resistance to UV light.
- Low specific gravity (1.25 g/mL), suitable for lightweight fabrics.
- It is readily melt spun.
- Low flammability and smoke generation.
- It can be dyed with standard PET disperse dyes and with deeper and brighter shades.
- It can be considered a 'sustainable' polymer in that it is made from renewable resources and is more readily biodegradable than many other polymers.

However, it has a low melting point at around 175°C in crystalline form and poor abrasion resistance. PLA can be draw textured on conventional false twist texturing machines.[11] However, heater temperatures in the primary zone typically need to be between 100°C to 140°C for a 2.0 m heater, depending on the yarn count and filament fineness, necessitating the use of electrical resistance heaters. Literature indicates that PLA filament yarns are sensitive to friction and temperature in the draw texturing process.[12] It is, however, possible that PLA with further developments, will find increasing applications in the draw texturing of yarns for apparel end-use.

13.5 References

1 CIRFS, *The Prospect for the World Markets for Textured Yarns to the Year 2010*, European Man-made Fibres Association.

2 Discussions with Oerlikon-Barmag, China, 2009.

3 www.barmag-oerlikon.com/Portaldata/1/Resources/barmag/pdf/barmag_eFK_brochure_en.pdf

4 D.K. Wilson, T. Kollu, The Production of Textured Yarns by Methods Other than the False Twist Technique, *Textile Progress*, 16, 3 (1987).

5 www.tx.ncsu.edu/jtatm/volume 3 issue 3/articles/ITMA/seyam_ITMA_full.pdf

6 *ITMA 2003*, www.allbusiness.com/asia/925832-1.html

7 Technical brochure, *TG.30 AE 'Combi'*, Giudici S.p.A., Italy.

8 Technical brochure, *False Twist Texturing, DP3-FT*, SSM Schaerer Schweiter Mettler AG, Switzerland.

9 S.S. Mahish, S.K. Laddha, PTT–Fiber of the Future, *Chemical Fibers International*, 300, 54, Oct. (2004).

10 J. S. Dugan, www.fibersource.com/f-tutor/pla.htm

11 *PLA Filament Texturing Process Guide*, www.natureworksllc.com/ProcessingGuides_FliamentTexturing_pdf

12 S. Kataoka, Muarata Machinery Ltd, JP 2001042720A, 20 Feb (2001).

Index